Structural Design of Air and Gas Ducts for Power Stations and Industrial Boiler Applications

Second Edition

Prepared by the Air and Gas Duct Structural Design Committee

ASCE

Published by the American Society of Civil Engineers

Library of Congress Cataloging-in-Publication Data

Names: American Society of Civil Engineers. Air and Gas Duct Structural Design Committee.
Title: Structural design of air and gas ducts for power stations and industrial boiler applications / Air and Gas Duct Structural Design Committee of the Energy Division of the American Society of Civil Engineers.
Description: Second edition. | Reston, Virginia : ASCE, American Society of Civil Engineers, 2020. | Includes bibliographical references and index. | Summary: "This report will assist structural engineers in understanding the structural behavior of a duct system and in analyzing and designing its many structural components"– Provided by publisher.
Identifiers: LCCN 2020007322 | ISBN 9780784415580 (paperback) | ISBN 9780784483008 (adobe pdf)
Subjects: LCSH: Power plants–Equipment and supplies. | Steam-boilers. | Air ducts–Design and construction. | Flue gases.
Classification: LCC TJ164 .S87 2020 | DDC 621.31/21–dc23
LC record available at https://lccn.loc.gov/2020007322

Published by the American Society of Civil Engineers
1801 Alexander Bell Drive
Reston, Virginia 20191-4382
www.asce.org/bookstore | ascelibrary.org

***Errata:* Errata, if any, can be found at https://doi.org/10.1061/9780784415580.**

ISBN 978-0-7844-1558-0 (print)
ISBN 978-0-7844-8300-8 (PDF)

Manufactured in the United States of America.
26 25 24 23 22 21 20 1 2 3 4 5

Contents

PREFACE

This ASCE publication has been created by a select committee of structural and mechanical engineers who are extremely experienced in the structural analysis and design of air and flue-gas ductwork for power stations and large industrial boiler applications.

The need for this ASCE publication was identified in 1991 by the ASCE Fossil Power Committee under the chairmanship of Thomas Longlais to capture industry practices for the design of air and gas ducts because they were not directly covered by building codes or other publications at the time. The original Air & Gas Duct Structural Design Committee was formed under Ron Schneider, and this committee finalized the original ASCE publication in 1995. Over 20 years later the original publication was useful, but had become dated because of new industry practices, code changes, and advancements in technology. The Air & Gas Duct Structural Design Committee was revived by the ASCE Energy Division in February 2016 to update the publication. A group of industry experts was assembled, and all 12 committee members were given the opportunity to review and comment on all 13 chapters of the publication. The resulting ASCE publication was unanimously agreed to by the committee.

Acknowledgments

This publication never could have been completed without the hard work and dedication of all the committee members. I thank all of them for donating their time and sharing their knowledge for the betterment of our industry.

Air and Gas Duct Structural Design Committee

Matthew Rosecrans, *Chair*	Sargent & Lundy
Kevin Como	Burns & McDonnell
Jason Fifarek	Sargent & Lundy
John Liu	Kiewit
Paul Nystedt	Babcock Power
Lucas Pachal	Nederveld, Inc.
Douglas Polack	American Electric Power
Dennis Richard	Black & Veatch (retired)
Dan Sack	Becht Engineering
David Six	Babcock & Wilcox
Eric Skibbe	Kiewit
Jeffrey Thompson	Consulting Engineer

Original Air and Gas Duct Structural Design Committee

Ron Schneider*	Gilbert/Commonwealth Engineers
Dan Blackwood	Southern Company Services
Vic Bochicchio	Zurn Balcke-Durr, Inc.
Bob Bucelwicz	Boston Edison Co.
Joe Clark*	ABB Combustion Engineering Systems
Roy Hogan*	ABB Environmental Systems
Bill Jacks	Tennessee Valley Authority
Ron Johnson*	Babcock & Wilcox
Paul Kokosinski*	Public Service Electric & Gas
Tim Laughlin*	Sargent & Lundy
Tom Longlais	Sargent & Lundy
Jim Newell	Stone & Webster Engineering Corp.
Rod Simonetti*	Gilbert/Commonwealth Engineers
Mike Stiefermann*	Central Electric Company
Ken Tamms	American Electric Power
Walt Van Dyke*	Foster Wheeler Energy Corp.
Ray Warren*	Warren Engineering
Heim Weinstein	Public Service Electric & Gas
Jim Whitcraft*	Bechtel Power Corp.

*Individuals of the original committee providing significant and time-consuming contributions to author various sections of this publication.

Other engineers and designers in the committee members' organizations also contributed to the development of this ASCE publication. Some significant contributors are

- Daniel Biss, David Wagner, Timothy Frymoyer, and Kenneth Bauer from Gilbert/Commonwealth Engineers;
- Kurt Wachholder and Edward Hanko from Sargent & Lundy; and
- Suzana Rufener from Babcock & Wilcox.

Matthew R. Rosecrans, P.E., S.E., M.ASCE

CHAPTER 1
Introduction

1.1 STATEMENT OF INTENT AND EXPECTED USE OF THIS BOOK

This ASCE publication has been created to assist structural engineers in the structural analysis and design of air and flue-gas ducts for power stations and industrial boiler applications. It is not intended to be used to size or configure ductwork for flow and pressure drop considerations.

The information in this book was written specifically for structural engineers. The ASCE committee responsible for creating this book strongly recommends that the structural analysis and design of ducts be performed by qualified structural engineers, and not by technicians, designers, or drafters.

The structural engineer is cautioned against reading selected sections without first reading and understanding the entire publication. Much of the information in each section depends on information in other sections. The structural engineer should read all the information presented in Chapters 1 through 13 and understand how it is interrelated before proceeding with analysis and design activities.

Air and gas ducts for power stations and industrial boiler applications are unique structures. The structural analysis and design of these ducts is currently not referenced or governed by any national code or design standard, and few published data on ductwork structural analysis and design procedures are available. Ductwork structural analysis and design is complicated by the need to accommodate large thermal movements and to assess the behavior of materials at high temperatures, sometimes in a corrosive or erosive environment. The structural behavior of steel ductwork can be difficult to understand for structural engineers inexperienced in ductwork analysis and design. Usually, ductwork is initially routed and configured considering only flow and pressure drop requirements, but the final arrangement must also consider structural support and structural behavior. In some organizations, ductwork design responsibility is divided between the mechanical process engineering and structural engineering disciplines. Often, problems that arise can be attributed to the failure to clearly assign and accept responsibilities between the disciplines. Communication between the disciplines and establishment of clear areas of responsibility are vital to ensure successful ductwork performance.

Ductwork arrangements have become more complex, and the design pressures have increased significantly, because of precipitators, fabric filters, NO_x removal systems (also known as selective catalytic reduction, or SCR), and scrubbers being added to the boiler exhaust system as flue-gas emission requirements became more stringent over the years. If the experienced structural engineer is not allowed input into the ductwork arrangement, a structural system could evolve that is prone to structural distress. Also, higher boiler outlet temperatures and more corrosive scrubber outlet gas chemistry have increased the importance of choosing the proper materials and the proper design strengths. There is a history in the power industry of failures and major degradation of flue-gas ductwork. There are many reasons for these failures or degradation, but many of the problems might have been avoided by better initial designs or material selection.

The information in this book may be used for international projects. However, if the materials to be used are different from those referenced in this book, special attention will be required from the structural engineer. Materials referenced by other international standards may be used and should be carefully assessed by the structural engineer, including mechanical properties, chemical composition, and material properties at elevated temperatures. Materials equivalent at ambient temperature may not have equivalent properties at elevated temperatures.

With this book, ASCE hopes to fill a void in the structural engineering field. It is expected to be used as a guide for structural engineers as they participate in the routing and configuration of a ductwork system, including the duct and its support structure. This book is also expected to be a valuable tool for structural engineers in understanding the structural behavior of a duct system and in analyzing and designing its many structural components.

1.2 LIMITATIONS AND SCOPE OF THIS BOOK

1.2.1 Scope

This book presents current approaches to the structural analysis and design of air and flue-gas ductwork. Included are sections on material selection, behavior, and performance; design loads, loading combinations, and design strengths; thermal considerations; vibration considerations; structural arrangement and behavior; toggle duct behavior and expansion joint considerations; overall duct structural analysis and design methods; and design considerations for local elements of the structure, such as stiffeners, internal braces, connections, turning vanes, and other flow distribution devices. It also discusses drawing and specification content, fabrication and construction techniques and considerations, duct support methods, and special considerations in the design of duct support structures. It discusses field maintenance examinations and inspections for preventative maintenance or condition assessment. Finally, it discusses the reinforcement of existing ductwork, recognizing that flue-gas emission requirements are becoming more stringent, requiring new environmental control systems that dictate modification and/or reinforcement of existing duct systems.

This book should be viewed differently from a design standard or code. It is broader in scope and does not necessarily endorse a single or specific design approach.

1.2.2 Limitations

This book is intended to be used for the structural analysis and design of ductwork where the metal temperature is below 1,100 °F (600 °C). The user is also cautioned that other criteria not addressed in this book may be appropriate for specific ductwork applications where the gas environment is extremely harsh chemically. It is intended to be used for the structural analysis and design of steel ductwork and is not necessarily applicable to ductwork of any nonmetal material, such as fiberglass reinforced plastic.

The book does not specifically address the structural analysis and design of fly ash collection hoppers, which are often part of ductwork systems. Hoppers are usually considered material storage bins, and their design requirements can be found elsewhere. However, in various sections this book discusses the interface of fly ash collection hoppers with ductwork.

Often, the structural engineer must address the vibration of ductwork structural elements. This consideration is especially important in duct sections immediately upstream and downstream of large fans in the duct system. This book addresses the need to perform designs that consider vibration concerns, but generally it is the responsibility of the structural engineer to take the proper actions so that excessive vibration does not occur. Vibration of local structural elements is discussed in Chapters 7, 8, and 9.

The book is not specifically intended to be used for the structural analysis and design of ductwork system equipment, such as the types listed in Section 1.5. However, the structural engineer performing the structural analysis and design of this equipment may choose to use any information presented in this book. The user is cautioned, however, to check the design of the equipment against any applicable design requirements or codes that have been published by others. For example, the structural design of precipitators is presented in Institute of Clean Air Companies (1993), EP-8, *Structural Design Criteria for Electrostatic Precipitator Casings.*

This book is also not intended to be used for the structural analysis and design of stacks, chimneys, or chimney liners. For the design of these structures, consult these publications as appropriate: American Society of Mechanical Engineers (ASME 2011, 2013), ASCE Task Committee on Steel Chimney Liners (1975), American Concrete Institute (2008), International Committee on Industrial Chimneys (CICIND) (2010), and CICIND publications *Model Codes for Concrete Chimneys* (https://cicind.org/publications/concrete-chimneys.html) and *Model Code for Steel Chimneys* (2010).

1.2.3 Application of Building Codes and Design Codes

This book is not to be interpreted as a design code for air and flue-gas ductwork, nor should it be interpreted to include all the restrictions and requirements presented in the various national, state, and local building codes and design codes, standards, and specifications. It is the responsibility of the structural engineer to review and consider the applicable design requirements, if any, presented in the governing building codes for the location of the project. This includes local building codes, which may have more stringent requirements than state and federal codes.

Traditionally, structural engineers have used the provisions of the American Institute of Steel Construction's *Specification for Structural Steel Buildings*, ANSI/AISC-360 (AISC 2017a), *Steel Construction Manual* (AISC 2017b), and *Minimum Design Loads and Associated Criteria for Buildings and Other Structures*, Standard ASCE/SEI 7-16 (ASCE 2017) when performing the structural design of air and flue-gas ductwork. Because the AISC 360 specification is intended to apply to conventional structures at ambient temperatures, the design of structural steel considering the effects of high temperatures has always been left to the judgment of the responsible structural engineer. This book anticipates the traditional use of the AISC 360 specification with adjustments and additional considerations for the high metal temperatures. In addition, the book provides structural analysis methods to predict the structural response at elevated temperatures.

This book has been prepared using US customary units, also known as standard English units (feet, pounds, and degrees Fahrenheit). There are also some empirical equations presented here that have relatively complicated constants, which are based on US customary units. Metric and Celsius units are given after the English units, in parentheses.

1.3 DUCTWORK SYSTEMS DESCRIPTIONS

For the purpose of this book, ducts are defined as airtight or gas-tight conduits that convey air or flue-gas under positive or negative pressure. They may or may not be exposed to high temperatures. Most ducts are circular or rectangular in cross section, but sometimes they may have an unusual shape as they transition into a piece of major equipment, such as a fan, precipitator, SCR, or scrubber. To accommodate temperature changes, ducts are usually independent structures that "float" on slide bearing plates or are suspended by a hanger system. Ducts typically connect pieces of major equipment but are not necessarily considered part of the equipment. The terms *ducts* and *ductwork* are usually used interchangeably. Ducts that convey flue-gas are also commonly called *flues* or *breeching*.

Given the ambiguity of the term *ductwork*, especially at the interface with major equipment, the user is responsible for defining the term and properly applying the information presented herein. Ductwork should not be defined by contract supply scope alone. For example, ductwork attached to a precipitator's inlet flange and supplied by the precipitator manufacturer should be considered ductwork, not part of the precipitator, but chambers that convey gas internal to a fabric filter are typically considered part of the fabric filter, not ductwork.

1.3.1 Ductwork Systems

Conventional Boiler Plants

There are two common types of ductwork systems in a conventional power plant or industrial boiler. They are called a *pressurized system* or a *balanced draft system*. In a large fossil fuel power plant, there are usually two distinct yet identical parallel ductwork streams each having its own duct system and set of identical equipment. This is done to ease fabrication and erection when a single duct may be too large and cumbersome to handle and install, and to allow the plant operators to isolate and perform maintenance on equipment without shutting down the unit. In smaller power plants or industrial boilers there is usually only one duct system and one piece of each equipment.

A *pressurized system*, also called a *forced draft system*, has forced draft (FD) fans upstream of the boiler that force air through the entire combustion system and eventually force the resulting flue-gas out to the atmosphere through the stack. Sometimes primary air (PA) fans are also in the system. These fans and their purpose in the combustion system are discussed in detail in Section 1.3.2. A pressurized system results in positive pressures in all the ducts. See Figure 1-1

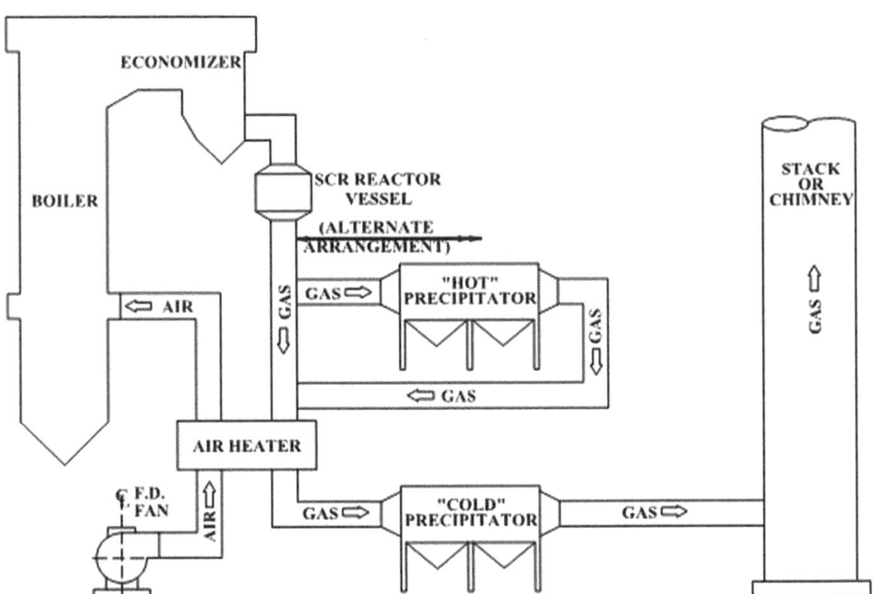

Figure 1-1. Typical pressurized system; ductwork arrangement for a coal-fired power plant.

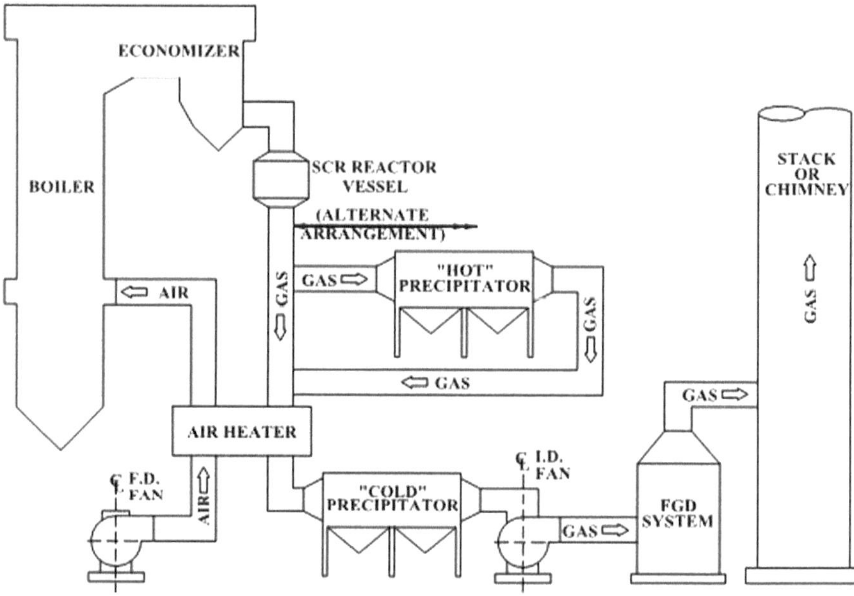

Figure 1-2. Typical balanced draft system; ductwork arrangement for a coal-fired power plant.

for a sketch of a typical pressurized system for a coal-fired power plant. The PA fans have been left off this sketch for simplicity.

In a balanced draft system, a combination of FD fans and induced draft (ID) fans are used. The FD fans are upstream of the boiler, and the ID fans are between the boiler and the stack. As with a pressurized system, PA fans may also be present. A balanced draft system results in positive pressures in the air ducts leading to the boiler and negative pressures in the flue-gas ducts between the boiler outlet and the ID fan inlet. The ductwork from the ID fan outlet to the stack is usually under a small positive pressure. See Figure 1-2 for a sketch of a typical balanced draft system for a coal-fired power plant. The PA fans have also been left off this sketch to keep it simple.

In any ductwork system there also may be various bypass ducts. Bypass ducts are used when the operators of the installation wish to, or must, shut off a major piece of equipment, such as an air preheater, SCR, or flue-gas desulfurization scrubber, on one or both sides of the unit. Bypass ducts are also used for reheating or tempering. In these cases, the operator is required to open and close various dampers to send the air or flue-gas through the bypass duct.

Combustion Turbine or Combined Cycle Power Plants

The ductwork system in a combustion turbine or combined cycle power plant is very different and much simpler than in a conventional coal power plant. This is because these plants burn natural gas as a fuel, and the exhaust flue-gas does not

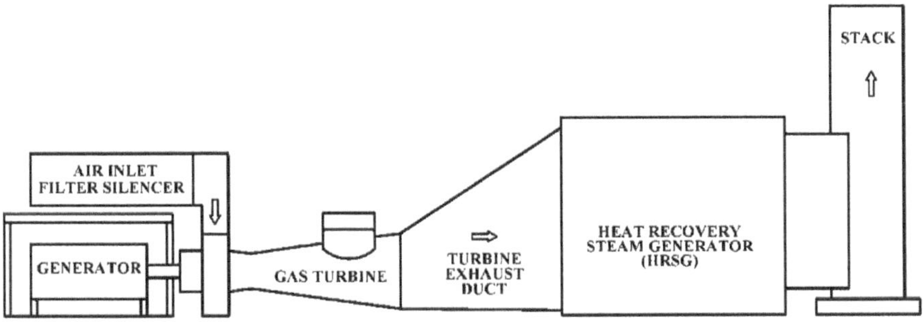

Figure 1-3. Typical ductwork arrangement for a combined cycle power plant.

require cleanup. For a combined cycle power plant, as shown in Figure 1-3, there are three sections of duct: the air inlet duct, the turbine exhaust duct, and the heat recovery steam generator (HRSG) outlet duct. The stack is usually nothing more than an extension of this outlet duct. The turbine exhaust duct is under a positive pressure and carries combustion exhaust flue-gas at temperatures in excess of 1,100 °F (600 °C). These ducts are usually internally insulated to protect the duct plate. The temperature in the HRSG outlet duct is typically around 450 °F (230 °C).

1.3.2 Air Ducts

Fluidized-Bed Boiler System

In a fluidized-bed boiler system, a combination of one PA fan and one FD fan supplies the combustion air to the boiler. In this system, the FD fan is sometimes called a secondary air fan. The primary air duct extends from the PA fan to the air preheater; downstream of the air preheater this duct is transformed into a manifold system that provides a uniform supply of fluidized air for combustion across the bottom of the furnace. The secondary air duct also extends from the FD fan to the air preheater. However, downstream of the air preheater this duct connects to the furnace section at a higher elevation to provide additional heated air for combustion. Figure 1-4 shows a typical air duct arrangement for a fluidized-bed boiler.

Conventional Boiler Systems

The ducts connecting the FD fans and the boiler convey the combustion air. These are called the *secondary air ducts*. The air usually goes through a preheater, which heats the air for more efficient combustion in the boiler. These ducts may also pass through a glycol or steam coil combustion air heater that heats the air to the temperature required by the air preheater manufacturer. The air is at ambient temperature upstream of the heaters and at a higher temperature, typically 200 °F to 500 °F (90 °C to 260 °C), between the air preheater and the boiler. These ducts are usually under a positive pressure.

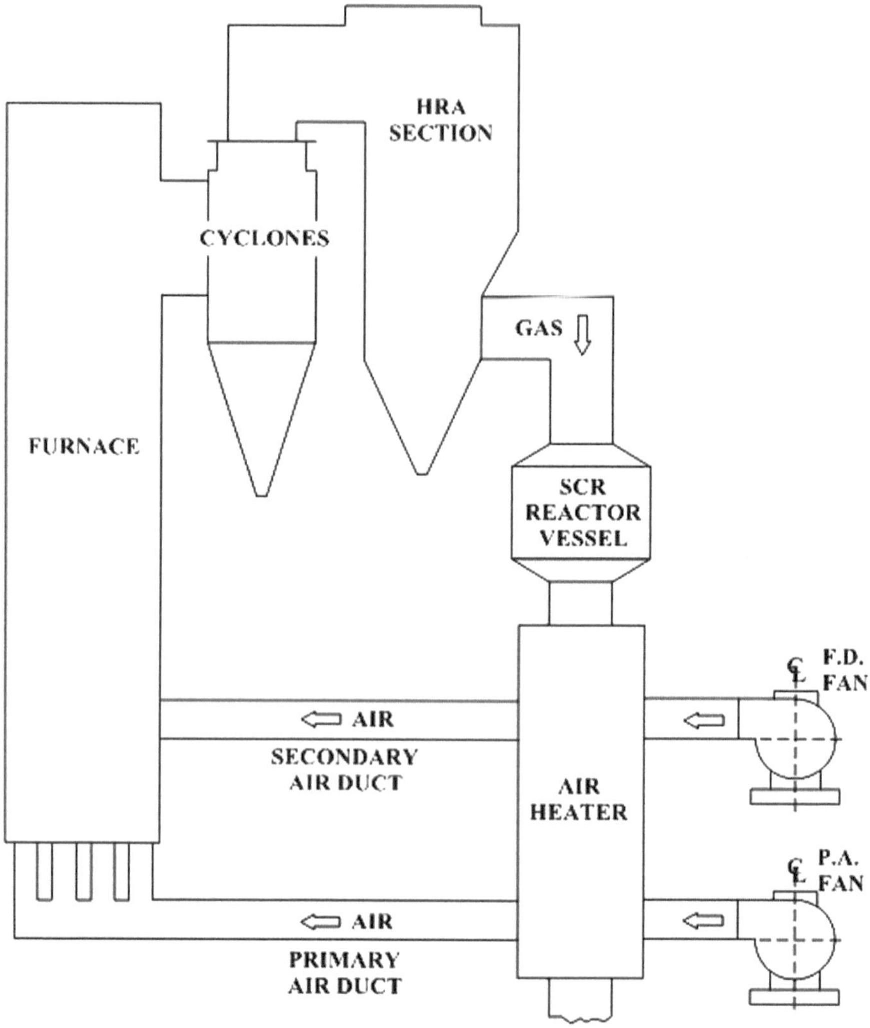

Figure 1-4. Typical air duct arrangement for a fluidized-bed boiler.

In a pulverized coal-fired boiler system, PA fans supply air, which conveys the pulverized coal from the pulverizers into the boiler. The ducts that supply the forced air from the outlets of the PA fans to the pulverizers are called the *primary air ducts*. The ducts connecting the PA fans to the pulverizers also go through an air preheater, and possibly a glycol or steam coil combustion air heater. They are also usually under a positive pressure and at approximately the same temperature as the secondary air ducts. In units that do not have PA fans, primary air is supplied from the secondary air ducts directly downstream of the air preheater.

Another type of air duct used in some pulverized coal-fired boiler systems is the *tempering air duct*. Its purpose is to convey ambient-temperature air to the

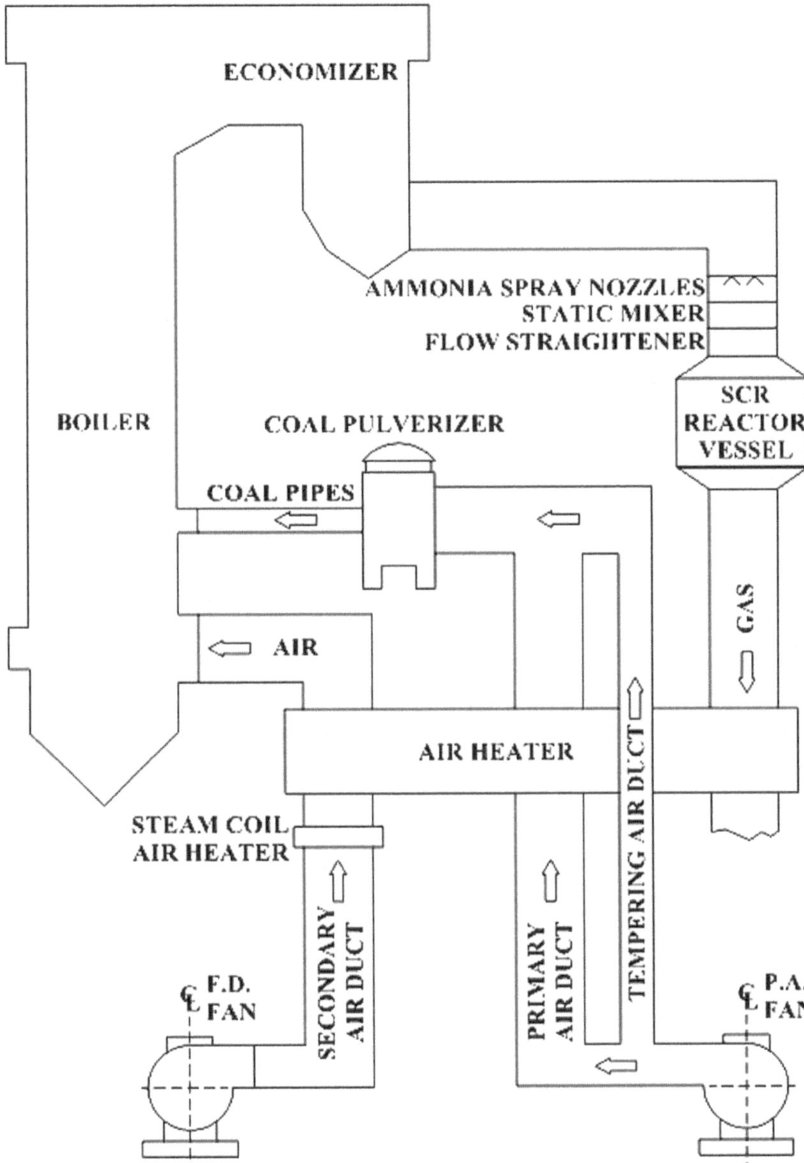

Figure 1-5. Typical air duct arrangement for a pulverized coal-fired boiler.

primary air duct mixing section. Tempering air is used to cool primary air to the proper temperature as it enters the pulverizers. Figure 1-5 shows a typical air duct arrangement in a pulverized coal-fired power plant. As shown in the figure, coal pipes or conduits transport the airborne coal from the pulverizers to the furnace. These coal pipes or conduits should not be considered ducts.

Other Air Ducts

Depending on the fuel, the application, the vendor, and the size of the boiler, there also may be other types of air ducts that are part of the duct system. For example, some boiler systems, such as recovery boilers, have tertiary fans and their associated air ducts. The air ducts discussed in this section are the most common ducts.

1.3.3 Flue-Gas Ducts or Flues

The flue-gas ductwork system between the boiler outlet and the chimney is usually more complicated. There can be many large pieces of equipment in the system. For a coal-fired boiler, the outlet flue-gas usually passes through an SCR, which reduces the oxides of nitrogen (NO_x); an air preheater, where it is cooled; an electrostatic precipitator or a fabric filter (baghouse), where most of the fly ash particles are removed from the flue-gas; the ID fans; and a flue-gas desulfurization (FDG) system, or scrubber, which removes sulfur.

Precipitators may be situated in the system in either of two locations: before the air preheater, on its hot side, or after the air preheater, on the cold side. Baghouses are always on the cold side. The ductwork system may or may not have a scrubber, depending on the age of the plant and the characteristics of the fuel burned. A description of these major pieces of equipment follows in Section 1.5.

The ducts from the boiler outlet to the air preheater convey hot flue-gas, at 600 °F to 900 °F (320 °C to 480 °C) or even hotter. In the air preheater, to increase plant efficiency, heat is transferred from the exiting flue-gas to the incoming combustion air. The flue-gas exiting the air preheater is cooled to approximately 250 °F to 450 °F (120 °C to 230 °C). If the system has a scrubber, the flue-gas typically will be cooled further, to between 120 °F and 350 °F (50 °C to 180 °C), as it passes through the scrubber. The structural engineer should understand that some systems have a scrubber bypass duct, which should be treated the same as the scrubber inlet ductwork.

Some power plants also have gas recirculation fans and their associated gas recirculation ducts. These ducts convey a portion of the combustion flue-gas back into the boiler furnace for steam temperature control and/or NO_x control. These ducts should be treated the same as the boiler outlet ducts.

Figures 1-1 and 1-2 show typical flue-gas duct arrangements in a pulverized coal-fired power plant.

1.3.4 Typical Pressures and Temperatures

As stated earlier, each section of ductwork in a system is exposed to a different pressure and possibly a different temperature. In a coal-fired power plant, the air is heated as it goes through the air preheaters, and the flue-gas temperature is cooled as it passes through the air preheaters and the scrubber. The pressure normally changes significantly as the flow passes through the equipment: the air preheater, the boiler, the SCR, the precipitator or baghouse, the scrubber, and all the major fans (Figure 1-6).

The structural engineer should obtain the proper design operating pressures from the responsible mechanical process engineer prior to starting the analysis

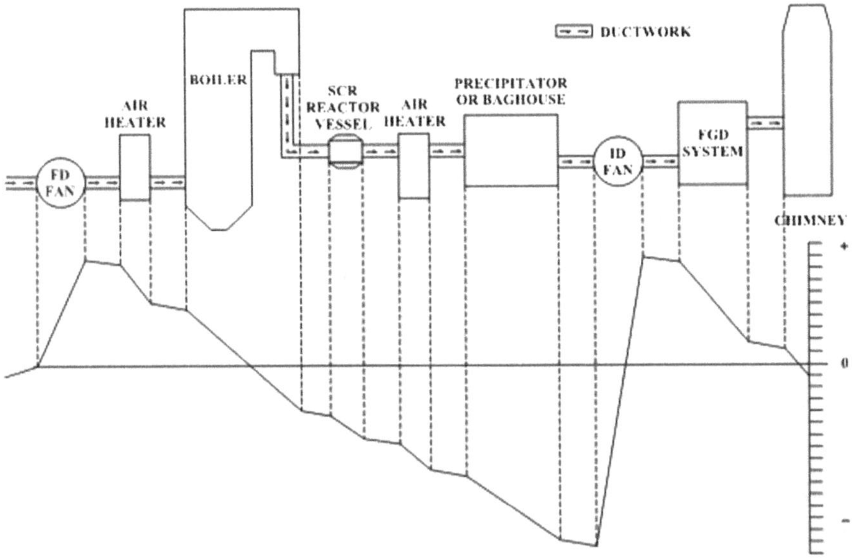

Figure 1-6. Typical pressure profile for a coal-fired power plant.

and design of the ductwork. The minimum design pressure requirements given in the National Fire Protection Association's *Boiler and Combustion Systems Hazards Code* (2015) may be applicable.

Representative air and flue-gas pressures and temperatures for a pressurized coal-fired power plant are shown as an example in Table 4-1. Typical representative values for a balanced draft coal-fired plant are shown as an example in Table 4-2. Typical values for pressurized and balanced draft industrial boilers are shown as an example in Table 4-3. The values in these tables are only examples. They should not blindly be used for ductwork design. The structural engineer should obtain the proper operating and transient pressures and operating and excursion temperatures for the duct system under consideration from the appropriate sources.

1.4 GLOSSARY/DEFINITIONS

Following are definitions of some common terms relating to ductwork and ductwork design that are used herein. The definitions are presented in six groups: ductwork, duct supports, duct appurtenances, internal flow distribution devices, duct service conditions, and duct conveyances.

1.4.1 Ductwork

Ducts, Ductwork: Ducts or ductwork can be any pressure-tight conveyance that transfers pressurized air or flue-gas from one point to another (see details in Section 1.3).

Flues: Ducts that convey flue-gas are sometimes called *flues*. Chimney liners and short square stacks used in natural gas burning plants are also sometimes called flues.

Hot-Side Ducts: Ducts or flues that convey the flue-gas from the boiler outlet or the economizer to the air preheater inlet are typically called *hot-side ducts*, just as precipitators located between the boiler outlet and the air preheater inlet are called *hot-side precipitators*. This definition arose because these ducts are on the hot side of the air preheater. The flue-gas temperatures in hot-side ducts are typically between 600 °F and 900 °F (330 °C and 480 °C). See Section 1.3.3 for more on flue-gas ducts.

Cold-Side Ducts: Ducts or flues that convey the flue-gas from the air preheater outlet through various pieces of major equipment to the stack or chimney are typically called *cold-side ducts*, just as precipitators located downstream of the air preheater are called *cold-side precipitators*. These flue-gas ducts are on the relatively cooler side of the air preheater. The flue-gas temperature in cold-side ducts is typically between 250 °F and 450 °F (120 °C and 230 °C). See Section 1.3.3 for more on flue-gas ducts.

Toggle Duct Section: A *toggle duct* is a straight section of ductwork between two expansion joints that is allowed to freely move in one, two, or three directions with one or two of the adjacent duct sections (or pieces of equipment), which are on the other side of an expansion joint. It is a section of ductwork that behaves as if it has a hinge at each end. See Figures 1-7 and 2-6 for sketches of a typical toggle duct arrangement. A toggle duct section typically does not have any external supports, although it may have one, usually at or near its center of expected rotation. A toggle duct is usually connected to the adjacent ducts across the expansion joints by shear ties or internal hanger rods, but carefully designed metal expansion joints may be used to transmit the calculated shear force. The purpose of a toggle duct section is to accommodate thermal movements, especially shear movements, that are too large for a single expansion joint. The most common location for a toggle duct section is at the boiler outlet. There, the boiler thermally expands down as much as 6 or 7 in. (15 or 18 cm). The toggle duct is connected to the boiler on one end and the main duct on the other end. The main duct is usually supported vertically near the expansion joint at the toggle section so that the vertical growth is close to zero. Therefore, the one end of the toggle duct goes down and up with the boiler, and the other end stays with the main duct. This same type of action also occurs in the horizontal plane. All rotation is taken out through the expansion joints. For more on toggle ducts, see Section 2.3.2.

Breeching: *Breeching* is the term commonly used for the section of flue-gas ductwork that enters the stack, chimney, or any piece of major equipment.

Hoppers: In ductwork applications, ash collection hoppers are located in sections of the ducts where significant dropout of fly ash is expected. Two typical locations for hoppers are at the boiler outlet ducts below the air preheater or economizer and at the bottom of vertical runs of ducts. The purpose of a hopper is to temporarily store the ash collected from the flue-gas flow until it is removed by the ash removal system.

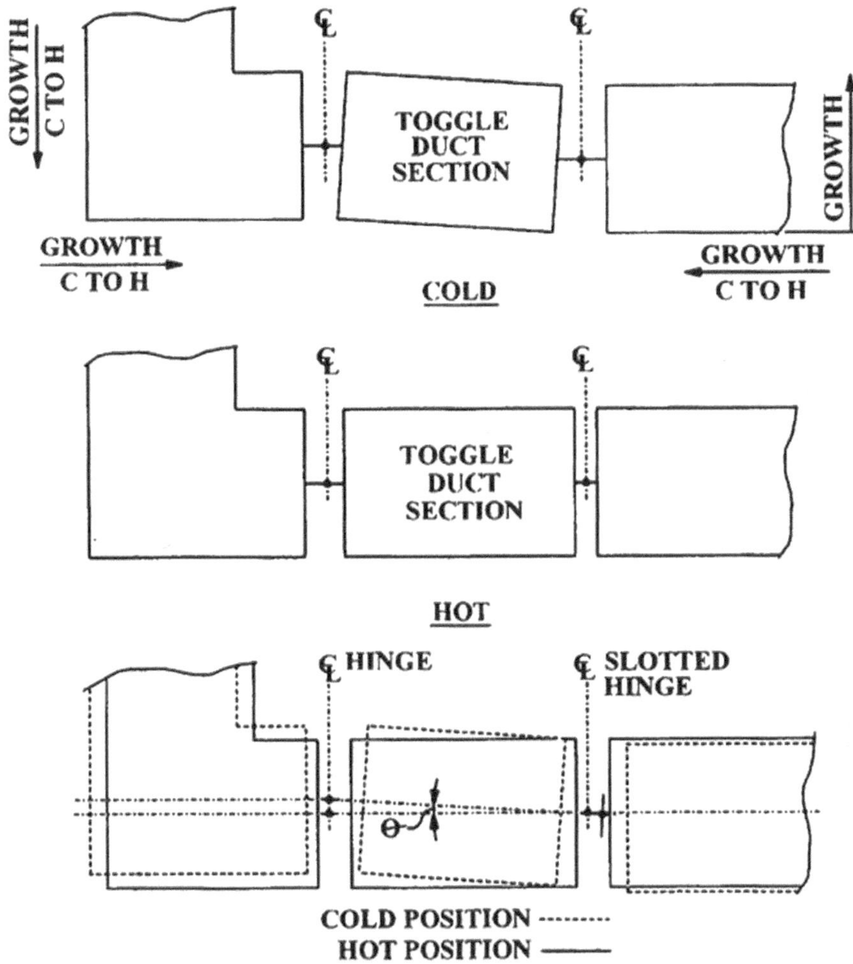

Figure 1-7. Typical toggle duct arrangement.

1.4.2 Duct Supports

Anchor Point: All ducts exposed to elevated temperatures experience thermal growth. Each section of duct must have a support configuration such that it may thermally grow in all directions without resistance. The location of zero movement is called the *anchor point* or *fixed point*. The anchor point does not have to be a gravity support. Also, an anchor point may be artificial, as further explained in Section 2.3. The intersection of two lines of duct guides that are typically perpendicular is the point of zero movement. This is considered the anchor point, even though there may be no physical attachment to the support structure at that location. The anchor point should be located to minimize the shear across expansion joints, to simplify the duct expansion behavior, to simplify the support structure arrangement, and to equalize the duct expansions.

Duct Guide: This is a connection between the duct and its support structure where the wind loads, seismic loads, expansion joint actuation forces, unbalanced pressure, and unbalanced friction loads acting perpendicular to one direction of thermal expansion are transferred from the duct to the support structure. Duct guides should be on one of the principal anchor lines of the duct section and must always allow thermal growth away from the anchor point. A duct guide may or may not be a gravity support for the duct. If it is a gravity support on a bottom-supported duct, a friction force is developed acting along the anchor line as the duct thermally expands. A slide bearing plate is usually under the gravity support baseplate to keep this friction force to a minimum.

Duct Sliding Supports: The gravity connections between the duct and its support structure, where unrestrained thermal growth is allowed, are called *sliding supports*. As the duct thermally grows, friction forces are developed acting away from the anchor point. A slide bearing plate is usually used under the gravity support base plates to keep these friction forces to a minimum.

Slide Bearing Plate Assembly: Slide bearing plates are typically used under all bottom-supported duct gravity support locations except the anchor point. There could be special cases where the gravity force is so low that a slide assembly is not cost-effective and therefore not used. The purpose of a slide bearing plate assembly is to reduce the friction between the duct support baseplate and the support steel. The friction typically is caused by the duct moving relative to the support steel with thermal expansion or contraction. In some instances, the friction can be caused by the support steel moving relative to the duct. A typical bearing assembly consists of a plate covered with low-friction material, such as polytetrafluoroethylene or lubricated bronze, that rubs against a polished stainless steel surface. The low-friction-material slide plate is usually attached to the fixed support structure, and the stainless steel plate to the moving duct support. Slide bearing plates are manufactured in flat, spherical, or a combined flat and spherical arrangement.

Constant-Support Spring Hangers: These types of hangers allow a predetermined vertical movement with a constant support resistance. They are used as vertical supports for duct sections at locations where the design engineer wants to control the load going to the support structure. They can be provided to take a local dead load, such as a damper, directly to the support structure to avoid having this load supported by the ductwork. As their name indicates, this type of hanger loads the support structure with a constant force, even if the forces or loads within the duct change. This type of hanger should only be used in conjunction with an accurate structural analysis. The dead load should be accurately determined, and the engineer should know where any additional load will go. To obtain the proper results, this hanger force should be applied as an external load in the duct analysis, not as a support.

Rigid Hangers: These hangers are usually rods pinned at both ends. They will act as a one-direction support, typically a vertical support, transmitting all loads to the support structure up to the capacity of the hanger with no vertical movement.

Variable-Support Spring Hangers: These hangers allow a predetermined vertical movement with a varying support resistance. They are used as vertical supports for duct sections at locations that will experience vertical thermal movement. They are used in locations where the structural engineer allows the load going to the support structure to vary.

1.4.3 Duct Appurtenances

Dampers: Dampers are placed in a duct system to allow the plant operator to control or redirect the flow within the duct system. Isolation dampers are also used to isolate sections of ductwork. There are five types of dampers that are generally used in ducts: *guillotine*, *louver*, *butterfly*, *diverter*, and *poppet*. A damper always comes with its own frame, which matches up with the duct so that it effectively becomes part of the duct. The location of dampers dictates the design and transient pressure forces acting on the duct and its support structure. Guillotine dampers are usually used for isolation because they generally are more leak-tight than multiblade louver dampers. Sometimes double-blade guillotine dampers are used where isolation is an absolute necessity. Guillotine dampers are often very heavy and require considerable space when the blade is out of the duct. To ensure complete isolation, pressurized-seal air frames are installed around dampers such that any leaks are always into the flue-gas stream instead of out to the atmosphere.

Expansion Joints: These types of joints are added to a ductwork system to divide the duct into sections to control thermal expansion and differential movements with high temperature. Expansion joints are also typically provided at the inlets and outlets of the major equipment in the system, such as the fans, boiler and economizer, SCR, air heaters, and scrubbers.

Expansion joints are made of either some sort of fabric or rubber-like material, or light-gauge metal (termed *nonmetallic expansion joints* and *metal expansion joints*, respectively). All expansion joints should be sized to accommodate duct axial movement and rotation. Some joints (usually nonmetallic) can allow a small amount of shear movement; others (usually metal) can be designed to transfer a shear force and allow no shear movement.

Insulation: Insulation is used on either the inside or the outside of ducts to keep the heat inside and to protect personnel from burns on the outside. Some older ducts were insulated on the inside with either masonry or mortar. Newer ducts are usually insulated on the outside with spun mineral fiber or fiberglass. If ducts are not insulated properly and there are cold spots, condensation will occur on the inside of the duct at these locations, and extensive corrosion of the steel will result. If hot-side ducts are not insulated properly, thermal shock could cause them to crack at or near the cold spots. See Chapter 11 for expanded discussion of ductwork insulation.

Lagging: This is the term for the fiberglass, aluminum, or light-gauge steel sheets that cover the ductwork insulation. Lagging keeps the insulation in place, protects the insulation from damage by weather or personnel, and keeps water off the insulation and the duct. See Chapter 11 for more on ductwork lagging.

Large Particle Ash (LPA) Screen: A screen is sometimes used to screen out large particles of ash that could clog equipment, such as the catalyst in SCR reactors. Such screens are typically situated to discharge the LPA into boiler outlet ductwork hoppers.

Corner Angle: Angles or bent plates typically are placed on the inside but sometimes are on the outside of ducts, in the corners between the wall plates and the floor and roof plates. Corner angles make erection easier, provide an airtight seal, and provide strength and stiffness to the roof, floor, and wall plates.

Internal Truss: Trusses are sometimes added to the inside of ducts to provide additional support to the roof and floor of very wide ducts; stiffen the walls of very tall ducts; and transmit wind, seismic, and pressure loads from the duct plate to the duct supports. The use of internal trusses is usually minimized, if possible, because they tend to constrict the flow and increase the pressure losses in the ducts. Their use should be evaluated by the structural engineer for duct economy and structural stiffness and by the mechanical engineer for flow considerations.

1.4.4 Internal Flow Distribution Devices

Perforated Plates: These are flow distribution devices within ducts, usually positioned perpendicular to the flow, that are sometimes necessary at the inlet and outlet of precipitators, baghouses, and other locations to ensure that all the chambers are receiving a uniform distribution of flow. A perforated plate is usually a flat, sometimes stiffened, plate with circular or slotted holes usually punched in a pattern. Perforated plates are described as having a certain percentage of open area, typically between 30% and 75%. Perforated plates are typically hung from the roof of the duct and are free from the floor and sides, so that they do not restrain the duct from moving unequally with any thermal expansion and contraction, wind deformation, pressure deformation, or gravity deformation. See Chapter 9 for the structural design of perforated plates.

Rectifiers: *Rectifiers* or *flow straighteners* are a series of cells, which may be square, circular, or in a honeycomb arrangement, and that are parallel to the flow to minimize flow disturbances. These are typically placed upstream of the SCR catalyst to provide a more even and less turbulent flow profile across the catalysts.

Splitter Plates: Plates such as these are internal flow distribution devices that are parallel to the flow. They are usually placed in a transition duct section to maintain an even flow within the duct and/or into a large piece of equipment, such as a precipitator. See Chapter 9 for the structural design of splitter plates.

Static Mixers: Static mixing plates are flow distribution devices that are used to mix reagents and reduce temperature, flue-gas, and fly ash stratification. They can also be used to prevent ash fallout in ductwork.

Turning Vanes: These types of vanes are internal flow distribution devices that help change the direction of flow while minimizing pressure drop. They direct the flow around duct corners, through duct transition pieces, and into major equipment. Turning vanes may be staggered flat plates called *ladder vanes*, or they may be parallel curved plates. Turning vanes are sometimes used in conjunction with splitter plates and perforated plates. See Chapter 9 for the structural design of turning vanes.

1.4.5 Duct Service Conditions

All these service conditions as they apply to loads are described in detail in Chapter 4.

Operating Pressure: Ducts usually are exposed to a relatively predictable static pressure range. Ducts may be pressurized, or under a partial vacuum. The static pressure is defined by the capabilities of the fans and the losses through the duct system, including the major equipment in the system. Each section of ductwork experiences a different static pressure, which is constant in that section during a given mode of operation. The maximum positive or minimum negative expected static pressure during normal operation is the operating, or design, pressure for the duct section. See Figure 1-6 for the typical operating pressure profile for a conventional coal-fired power plant. Pressures are normally specified in inches of water column or millimeters of mercury.

Unbalanced Pressure: Whenever a section of duct has an expansion joint located so that there is a duct wall that does not have another duct wall directly across from it, an unbalanced pressure occurs in that duct section. The resulting unbalanced pressure force must be transferred through the duct to its support structure. If no mechanism is available to resist the unbalanced pressure force, it will cause the duct section to move off its supports, which could cause considerable damage to the duct, its support structure, and nearby structures and equipment. Unbalanced pressure forces will only occur opposite openings or expansion joints directly upstream or downstream of bends or transitions. The unbalanced operating pressure force is the static operating pressure multiplied by the area of the opening. The structural engineer may often be able to exercise some level of control over these forces through judiciously orienting openings, improving the geometry of transition sections, keeping the duct runs as straight as possible, and ensuring the proper placement of dampers and expansion joints. This unbalanced force should be accounted for in the design of the ductwork supports and the support structure, as shown in Figure 4-2.

Dynamic Unbalanced Pressure: Dynamic pressure is associated with the velocity and density of the flow. This pressure fluctuates with time and flow rate. Given the low density of air and flue-gas, dynamic unbalanced pressure is normally only considered when designing flow distribution devices and sometimes for duct walls at turns. See Chapters 4 and 9 for discussions of dynamic unbalanced pressure.

Transient Pressure: During an abnormal event, ducts may experience a very high or very low static pressure for a relatively short period. Examples of such an event are the failure of a major piece of equipment, such as an ID fan; the improper closing of a damper; or a master fuel trip of the unit. This abnormally high (positive) or low (negative) pressure is defined as the *transient pressure*. Ducts should be designed considering the likelihood of experiencing the transient pressure, but a lower factor of safety may be used because of the low frequency and low chance of occurrence and the expected brief duration of the pressure spike.

Operating Temperature: Ducts usually are exposed to a relatively predictable temperature range, as governed by the boiler and air preheater design. The maximum expected temperature during normal operation is the operating or design temperature for each section of duct.

Excursion Temperature: During an abnormal event, such as the failure of a major piece of equipment (such as an air preheater) to perform properly, the ducts may experience a very high temperature for a relatively short period. This is the excursion temperature. Ducts should be designed considering the likelihood of experiencing the excursion temperature, but a lower factor of safety may be justified because of the low frequency and low chance of occurrence and the expected brief duration of the temperature rise.

Thermal Gradient: With unusual or unexpected flow, ducts sometimes experience uneven temperatures from top to bottom and/or from side to side. This condition results in thermal gradients within the duct, which should be considered in the analysis and design if they are known to be large or if they are expected to create significant stresses.

1.4.6 Duct Conveyances

Primary Air: This is the part of combustion air that is mixed directly with the coal before ignition takes place.

Secondary Air: This is is the part of combustion air that is added shortly after ignition to help complete the combustion process.

Tempering Air: This is air that is used to control temperature, especially to cool primary air to the proper temperature as it enters the pulverizers.

Tertiary Air: This air is that part of combustion air that is added to the burning mixture to complete the combustion process and to further mix the air and the coal.

Flue-Gas: This is the exhaust from the combustion process that takes place in the boiler. Flue-gas is conveyed from the boiler to the stack via ducts.

Recirculated Gas: This kind of gas is flue-gas that is fed back into the combustion zone for steam temperature control and/or NO_x control.

Fly Ash: This is the term for the nongaseous products of solid or liquid fuel combustion that are entrained in the flue-gas stream. Most fly ash is extracted from the flue-gas by the air pollution control equipment, usually precipitators or fabric filters. Fly ash also tends to fall out of the flue-gas and accumulate in the ducts in locations where the flow velocity is reduced and at turns and obstructions in the duct.

Salt Cake: This is the term for the ash from a black liquor (paper mill byproduct) recovery boiler.

Scrubber Sludge: *Sludge* is the term for the by-product of the wet flue-gas desulfurization process. Sludge is collected in the scrubber. Scrubber outlet ducts sometimes have an accumulation of sludge on their internal surfaces.

1.5 DESCRIPTIONS OF MAJOR DUCTWORK EQUIPMENT

Following is a list of the major equipment that can be involved with a ductwork system. Not every piece of equipment listed here will be found in a single duct system. The ductwork is typically designed around the requirements of this equipment. Also included is a short description of their function in the system and how the equipment interfaces with the ductwork. This equipment is typically not designed with the ductwork but purchased from specialty manufacturers.

Air Preheater: This is a device that transfers heat from the exiting flue-gas to the incoming combustion air. This process promotes more efficient combustion in the boiler and sends less unused energy up the chimney. There are many types of air preheaters, including regenerative circulating air preheaters, tubular air preheaters, and heat pipes. There are large pressure losses through most air preheaters.

Ash Removal Equipment: For a flue-gas ductwork system to operate properly as designed, all ash collection hoppers should be periodically emptied. Various ash removal systems may be used, but a common type is a vacuum removal system, where ash removal piping is attached to the bottom of the hoppers and the collected ash is intermittently vacuumed out. A continuous mechanical ash removal system has become the frequent choice in boiler systems that burn one of the many high-ash fuels, such as waste wood or refuse. Often the ash removal system is supported directly from the hoppers, and the additional weight should be considered in the structural analysis and design.

Boiler or Furnace: This is where the fuel is burned and water heated in pressurized tubes to create steam. The boiler is not typically considered ductwork except for the outlet casing downstream of the tubes.

Combustion Air Heater: These heaters heat the air before it enters the air preheater. Combustion air heaters usually use either glycol or steam coils to heat the air. Not all systems have combustion air heaters.

Continuous Emission Monitoring System: Flue-gas continuous emission monitoring was mandated by the 1991 Clean Air Act. A continuous emission monitoring system (CEMS) collects and analyzes the flue-gas. The concentrations of pollutants, NO_x and SO_2, plus diluents, CO_2 or O_2, are measured. The amount of flue-gas flow is also measured. These data are reported to state, local, and federal agencies. High levels of accuracy and availability are required for this equipment.

Economizer: An economizer is located at the rear of a boiler and uses the furnace exit flue-gas to preheat the de-aerated boiler feedwater.

Heat Recovery Steam Generator: An HRSG is a relatively large heat transfer device downstream of the gas turbine in the ductwork system of a combined cycle power plant. It is composed of groups of water-filled tubes that are heated by the turbine exhaust flue-gas. In an HRSG, the temperature of the flue-gas is considerably reduced, and the water in the tubes is converted to steam. This steam can then be used for a commercial purpose or drive a companion steam turbine.

Air Pollution Control Device: This is a term for mechanical ash collectors, precipitators, SCRs, fabric filters (baghouses), flue-gas desulfurization systems (scrubbers), and similar pieces of equipment that are designed to remove pollution particles from the flue-gas.

Electrostatic Precipitator: These remove most of the fly ash from the flue-gas using electrostatic energy. Within the precipitator, the flue-gas passes between parallel, vertically hanging ash collecting plates. Between the plates are electrically charged wire electrodes that induce an electric charge on the fly ash. This arrangement causes the fly ash to be attracted onto the collecting plates. The collecting plates are periodically rapped to dislodge the ash and allow it to fall into large collecting hoppers at the bottom of the precipitator before it is removed by the ash removal system. There are usually large pressure losses as the flue-gas passes through a precipitator and its transition ductwork.

Fabric Filter or Baghouse:

These also remove most of the fly ash from the flue-gas. The flue-gas is pulled through fabric bags, which act as filters to catch the fly ash. The bags are periodically emptied into collection hoppers underneath the fabric filter. Large ID fans are typically required if a fabric filter is in a ductwork system because of the large pressure losses as the flue-gas passes through the filter bags.

Flue-Gas Conditioning: Flue-gas is sometimes conditioned by injecting chemicals into the flue-gas stream downstream of the boiler or economizer outlet. One of the processes is selective catalytic reduction (SCR), which reduces the oxides of nitrogen (NO_x) in the flue-gas by injecting ammonia into the flue-gas stream in the presence of a catalyst. Another process involves injecting SO_3 into the flue-gas stream. These processes enhance the collection efficiency of precipitators in systems that use low-sulfur coal and thereby reduce pollution.

Flue-Gas Desulfurization (FGD) System or Scrubber: This system removes sulfur dioxide (SO_2) from the flue-gas using limestone, lime, or other reagent. Such systems may be wet or dry. There may be large pressure losses and a significant reduction in temperature as the flue-gas passes through the scrubber system.

Forced Draft (FD) Fan: These are in the ductwork system upstream of the boiler. They take in air from the atmosphere or the boiler enclosure and force it into the boiler for combustion. Depending on the size and type of boiler, there may be one to four FD fans associated with a combustion system.

Gas Recirculation Fan (GRF): Some power stations also have GRFs. These fans return a portion of the combustion flue-gas from the boiler outlet to the boiler furnace for steam temperature control and to control the NO_x formation. If any, there are usually one or two GRFs associated with a coal-fired combustion system.

Induced Draft (ID) Fan: These are in the ductwork system downstream of the boiler, between the precipitator or baghouse and the chimney. They pull the flue-gas out of the boiler and through the precipitator or baghouse and push it through the scrubber, if present, and out to the atmosphere through the chimney or stack. Depending on the size and type of boiler, there may be one to four ID fans associated with a combustion system.

Primary Air (PA) Fan: These are also upstream of the boiler, usually adjacent to the FD fans. They take in air from the atmosphere or the FD fan discharge and force it through the coal pulverizers, where the pulverized coal is entrained, and into the boiler for burning. They are usually only used with coal-fired boilers. There are usually one or two PA fans associated with a coal-fired combustion system.

Other types of fans: There may be fans in duct systems other than those mentioned here. Some others are the secondary air fan, which is used in the secondary air duct; tertiary air fan, used in the tertiary air duct; reheat air fan, used in a reheat duct; and reverse air fan, used to move reverse air within a fabric filter. All these types of fans generally have the same pressure-changing impact on the duct design.

Selective Catalytic Reduction (SCR): The SCR unit is typically located between the boiler economizer and the preheater. It converts the nitrogen oxides (NO_x) from the flue-gas into diatomic nitrogen (N_2) and water (H_2O) by adding anhydrous ammonia, aqueous ammonia, or urea to the gas stream and the reagent chemical and the nitrogen oxides are absorbed onto a catalyst. There may be pressure losses as the flue-gas passes through the SCR system.

Stack or Chimney Liner: This is the orifice at the end of the ductwork system through which the flue-gas enters the atmosphere. In common usage, *stack* and *chimney* are interchangeable. Generally, *chimney* refers to a concrete windscreen structure that protects a liner or liners that convey the flue-gas. Sometimes these liners are also called flues. A concrete chimney is typically cost-effective when the required flue-gas exit height is above 200 to 300 ft (60 m to 90 m), depending on the diameter. An unprotected carbon steel liner or stack cannot usually be used in conjunction with a scrubber because of the corrosive flue-gas exiting the scrubber. Brick or fiberglass reinforced plastic chimney liners are usually used in conjunction with scrubbers, although high-alloy metal and special acid-resisting coatings could also be provided.

1.6 DUCTWORK ACCESSORIES

Following is a list of accessories that are typically supplied with the ductwork. They are most often designed along with the ductwork and/or affect the design of the ductwork. Also included is a short description of their function in the ductwork system and how they interface with the ductwork.

Access Doors: These are usually installed in duct walls so that the ducts can be entered for maintenance when the boiler is out of service. The doors should be large enough for relatively easy entry but not so large that they weaken the duct wall. They should be designed to withstand the same pressures and wind loads as the duct wall. Door hinges and latches, or bolts, should also be sized for these same loads. Doors should be airtight. This is typically accomplished by providing a gasket seal around the entire door perimeter. Doors should also be properly insulated and lagged to prevent cold spots so that they are not damaged by corrosion.

Gaskets: Access doors should be sealed with an airtight gasket. The gasket should be chosen considering the design pressure and temperature and the flow chemistry. Sometimes gaskets are also used where a duct section connects to equipment, expansion joints, and dampers.

Instrumentation and Test Ports: Flanged pipe pieces typically penetrate the duct roof or walls at various locations so that air or gas measurement equipment can be inserted into the stream. Some of the instrumentation is permanently attached to the ports, but some of the pipe pieces are capped so that temporary or portable instruments can be inserted at any time. The locations of the ports are set by regulatory requirements, duct configuration, test data, operating needs, instrument manufacturer recommendations, and major equipment manufacturer recommendations.

Internal Guardrail: Sometimes it is necessary to install a guardrail within ducts to prevent personnel from falling off the floor of a horizontal section into the shaft of an adjacent vertical section.

Platforms and Ladders: Sometimes access platforms and ladders are supported directly off ductwork. If this is done, care must be taken in the structural design. The differential thermal expansion between the hotter duct and the cooler ladder or platform support steel should be accounted for when connecting the ladder or steel to the duct.

References

ACI (American Concrete Institute). 2008. *Code requirements for reinforced concrete chimneys and commentary*. ACI 307. Farmington Hills, MI: ACI.

AISC. 2017a. *Specification for structural steel buildings*. AISC 360-16. Chicago: AISC.

AISC. 2017b. *Steel construction manual*. AISC 325-16. Chicago: AISC.

ASCE. 2017. *Minimum design loads and associated criteria for building and other structures*. ASCE 7-16. New York: ASCE.

ASCE Task Committee on Steel Chimney Liners. 1975. *Design and construction of steel chimney liners*. New York: ASCE.

ASME. 2011. *Steel stacks*. ASME STS-1. New York: ASME.

ASME. 2013. *Reinforced thermoset plastic corrosion resistant equipment*. ASME RTP-1. New York: ASME.

CICIND (International Committee on Industrial Chimneys). 2010. *Model code for steel chimneys: The CICIND chimney standard*. Brighton, UK: CICIND.

Institute of Clean Air Companies. 1993. *Structural design criteria for electrostatic precipitator casings: Publication EP-8*. Arlington, VA: Institute of Clean Air Companies.

National Fire Protection Association. 2015. *Boiler and combustion systems hazards code: NFPA 85*. Quincy, MA: National Fire Protection Association.

CHAPTER 2

Ductwork Arrangement and Behavior

2.1 OVERVIEW

This section discusses how decisions made in the layout of a duct system may affect factors such as initial cost, structural behavior, system reliability, and maintenance. The design of large air and gas ducts is an extremely complex task that requires the satisfaction of both mechanical and structural criteria. Structural engineers should remember that the purpose of ducts is to efficiently convey the air or flue-gas from one location to another while, as much as possible, maintaining the temperature and pressure. The combustion process and air quality control considerations are just as important as the structural aspects addressed in this book. Flow considerations, such as minimizing pressure drop and controlling ash deposition, are also essential to a good overall design. The ability to accurately monitor emission levels, equipment operating conditions, and ease of maintenance are critical to the customer. Therefore, it is imperative that the structural engineer work closely with the other disciplines to coordinate the various design inputs and optimize the final overall duct design. This teamwork is especially important in the early stages of a project, when the duct and equipment arrangements are being developed.

2.1.1 Flow Considerations

Flow-related phenomena often have a large impact on both the initial cost and the operating cost related to ducts. For example, significant fly ash deposition caused by low flue-gas velocities or poor flow distribution will require a heavier duct and support structure to support the high ash loads. Also, an greater pressure drop will require more auxiliary power to run the fans. Flow model tests can predict pressure drops, flow stratification, thermal gradients, ash deposition, and liquid droplet re-entrainment. Flow model tests can also be used to help optimize duct layouts and the arrangement of flow distribution devices such as turning vanes, splitter plates, mixing plates, and perforated plates. If flow model tests are performed, the structural engineer should actively participate in the process so that structural design, internal structural elements, and constructability issues are addressed.

2.1.2 Design Process

The arrangement of a ductwork system, like any other engineered structure, should be performed with the goals of minimizing initial cost, optimizing overall constructability, optimizing system performance, and minimizing future maintenance costs. In some cases, these four general criteria influence design decisions similarly. For example, taking the most direct path between two points in the system is likely to result in both the most economical duct and the smallest pressure drop, because its length is minimized and turns have been avoided. In other cases, the design team must compromise between conflicting criteria. An example of this would be when the mechanical process engineer suggests using rounded corners, which would enhance the flow characteristics but may complicate the structural behavior and construction details.

The ductwork design process may vary considerably from individual to individual or company to company, but a typical design approach might have the following steps.

1. A mechanical process engineer prepares process flow and instrumentation diagrams of the duct system, which indicate the path(s) the air and flue-gas must follow from their entry into the system, through the various pieces of equipment, and out the chimney or stack.
2. The mechanical process engineer reviews the combustion and air quality control processes and identifies the flow rates, pressures, and temperatures in each section of ductwork. All transients and excursion pressures and temperatures are to be included in this determination, along with duct materials and erosion and corrosion allowances. Duct materials and erosion and corrosion allowances are selected with multidisciplinary input from mechanical, structural, and materials application specialists. Some projects may request owner input for the material selection and erosion and corrosion allowances.
3. Physical arrangements of the duct system are developed based on actual or estimated equipment sizes. Initial duct sizing is usually based on maintaining the velocities in the various duct sections within an acceptable range; these initial sizes are then modified to take into account other factors such as space constraints, the flange size at equipment connections, and possible fabrication, shipping, and erection constraints. Routing of the duct segments from point to point within the system is influenced by space constraints, flow considerations, equipment criteria, available means of support, and cost. Changes in duct cross section and changes in direction are typically minimized because of their effect on flow stratification, turbulence, and pressure drop. Expansion joints, dampers, supports, access doors, flow distribution devices, and internal structural elements are located by the project team. This step is often an iterative process involving both the structural and mechanical process engineers. Instrumentation and equipment to monitor flow, performance, and pollutant levels should also

be defined and located at this stage. The support structure and access arrangement should be developed in parallel with this stage of the duct arrangement.

4. Physical and/or computational flow model tests are often performed at this stage to verify performance assumptions, finalize pressure and flow criteria, and help determine flow distribution device configurations. The structural engineer should provide preliminary internal elements to the flow modeler so that these items are addressed. This includes typical turning vane details for the plate, stiffeners, and support structure. Damper blade and expansion joint flow liner geometries should also be provided.
5. The structural engineer selects the appropriate support types and identifies the locations to be anchored and guided. This effort should be closely coordinated with the location of the expansion joints. Thermal movements of each duct section are usually calculated at this point and used as input to the duct analysis and the expansion joint procurement.
6. The global and local structural analysis and design of the duct sections and their elements is performed by the structural engineer. Final arrangements and sizes for internal stiffeners, trusses, struts, and flow distribution devices should be reviewed by the mechanical process engineer for their effect on flow and pressure drop.
7. The structural analysis and design of duct support structures and their foundations is performed by the structural engineer using the loads and eccentricities resulting from the duct analyses.

2.2 INTERFACES WITH EQUIPMENT

Ducts must interface with many different types of equipment that contribute to the combustion and air quality control processes. Some equipment, such as an electrostatic precipitator or a fan, is external to the ductwork. In these cases, the duct sections connect to a mating flange at the equipment inlet and outlet. Other equipment, such as a steam coil air preheater or a flue-gas conditioning system, may be located inside the ducts. This section will address important aspects of the interface between ducts and equipment. See Section 1.5 for a brief description of the most commonly encountered ductwork equipment.

2.2.1 Isolation of Major Equipment from the Ductwork

Generally, ductwork should be structurally isolated from major equipment such as fans and air quality control devices. Isolation serves the following purposes:

- Simplifies the structural analysis of the components of the system by preventing the transfer of loads to and from the equipment across the isolation joint, making the structural behavior more predictable;

- Improves the ability of duct system components to expand and contract under temperature changes;
- Reduces the transmission of vibrations from the equipment to the ducts; and
- Establishes a clear division of responsibility between the equipment and duct suppliers.

Equipment isolation is usually accomplished through the placement of an expansion joint at the duct interface. Nonmetallic expansion joints provide the best isolation because of their low, often negligible, stiffnesses in all directions. Metal expansion joints may be used if isolation is only required in the axial direction, although the greater stiffness of a metal joint will reduce its effectiveness as an isolator. Because of their greater stiffness, metal joints tend to transmit more vibrations, create greater thermal restraint, and develop more indeterminate load paths in the ductwork than do nonmetallic joints. Therefore, the potential for transferring loads or vibrations to and from equipment should be investigated by the structural engineer when metal expansion joints are used. Nonmetallic joints may transmit vibrations if they "flutter" in the flow; however, this kind of fluttering usually has little effect on the ductwork design. Conversely fluttering of any expansion joint should not be tolerated, because the joint will eventually fail if the flutter is not eliminated.

2.2.2 Fans

Many types of fans can be present in a ductwork system, as discussed in Section 1.5. When arranging the ductwork adjacent to a fan, the performance of the fan and the effects of the fan on the duct are to be considered. The fan manufacturer may have minimum straight duct length requirements or restrictions on the shape of transition sections on either side of the fan. The fan manufacturer may also recommend minimum distances between the fan and internal duct elements such as structural trusses, struts, or flow distribution devices or require a specific flow distribution profile to ensure adequate fan performance.

In the absence of specific criteria from the fan manufacturer, the design team should consider providing a constant duct section at the fan outlet with no internal elements for a length equal to the "100% effective duct length" as defined in Section 8 of the Air Movement and Control Association's *Fans and Systems* (2007). For centrifugal fans, it is equal to 2-1/2 duct equivalent diameters for a flow velocity of 2,500 fpm (feet per minute) or less plus 1 duct equivalent diameter for each additional 1,000 fpm; for axial fans it is 50% of that required for centrifugal fans. The equivalent diameter may be taken as 1.13 times the square root of a rectangular duct's area. Internal duct elements should be avoided within this length of duct. Failure to meet criteria such as this could reduce the performance of the fan or could be cause for voiding of the fan warranty. Structural damage to the fan or duct from flow-related vibrations may also result.

2.2.3 Equipment Removal

The layout and design of ductwork systems should often consider the ability of the owner to maintain equipment around or within the ducts. The arrangement should ensure that the ducts and their supporting structures do not infringe on the space needed to replace or maintain nearby equipment. Removable panels may need to be provided in the duct walls to allow the removal of large pieces of equipment, such as fan wheels; access hatches may be needed to replace mechanical systems inside the ducts, such as the catalyst bed of an SCR system. Trolley beams may be supported from the ducts or their supporting structures to assist with equipment maintenance or removal. All these factors should be reviewed by the design team and the customer prior before the ductwork system layout is finalized.

2.2.4 Dampers

Dampers are an essential part of most duct systems, because they control the flow of air or flue-gas through the ducts. Most dampers function as shut-off dampers and, in any given operating scenario, are either fully open or fully closed. But some are installed to modulate the gas flow and therefore function in various partially open positions.

Figure 2-1 shows the three most common types of dampers: louver, butterfly, and guillotine. There are also other types less commonly used, such as poppet and diverter. Guillotine dampers usually have a single blade that slides in and out of the flow path, although some large dampers may have multiple or opposed blades. Guillotine dampers are almost always used as shut-off dampers, and they generally function best if the blades are kept in a vertical plane. Louver dampers have one or more rotating blades and may be used as either shut-off or flow-regulating dampers.

When dampers are located next to expansion joints, it is usually better to place the damper on the upstream side of the joint (Figure 2-2). This arrangement may increase the useful service life of the expansion joint by reducing its exposure to the flow environment. Also, and perhaps most important for ducts carrying air or flue-gas at elevated temperatures, in this arrangement the damper will be hotter than the downstream, inactive duct section. Placing the damper upstream of the expansion joint allows the hot damper to be structurally isolated from the cold duct, reducing the potential for high stresses in the duct and damper frame caused by the difference in temperatures. Caution should be used if an expansion joint is not installed adjacent to a damper. The damper supplier should be informed of the installation arrangement and all operating and excursion design conditions to achieve a reliable design.

Dampers are often quite heavy. The bonnets of guillotine dampers also represent large surface areas for wind load and drifting snow load. Because of these loads, the structural system will usually be most efficient if dampers are located as close to the duct supports as possible. Also, unless the damper manufacturer has been specifically instructed that the damper is to be an integral part of the

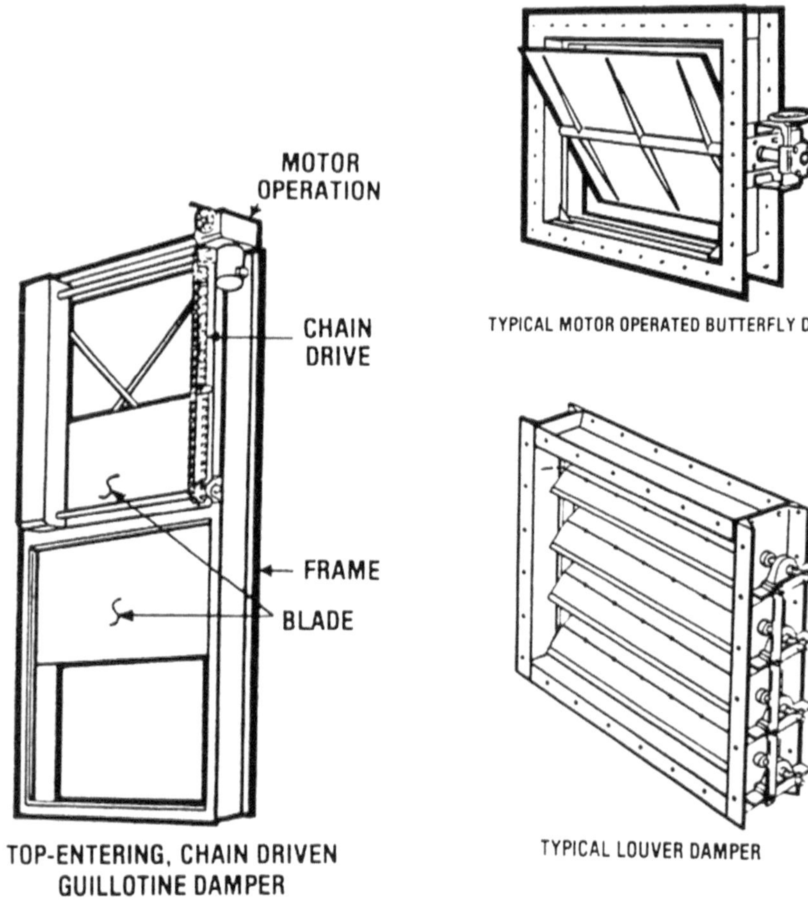

Figure 2-1. Damper types.

structural system of the duct, the damper frame is unlikely to have enough strength or rigidity to transmit very large shears and moments. Warping of the damper frame, as it tries to transfer loads and elastically deflects with the duct, may impair the functioning of the damper. Therefore, dampers should be located in regions of low stress, such as adjacent to expansion joints. For similar reasons, ductwork should not be supported from dampers. If the damper's location within the duct arrangement causes the damper to transmit or carry loads, the damper manufacturer should be advised and confirm acceptability. Whenever a damper is located adjacent to an expansion joint, enough clearance is to be provided so that the expansion joint material, including flow liners, does not interfere with the damper blade movement. In addition, clearance between the damper blade sweep and internal elements (trusses, turning vanes, etc.) should be checked.

In general, shut-off dampers should be located in such a way that dead legs are minimized. As shown in Figure 2-3, dead legs are sections of duct that are open to

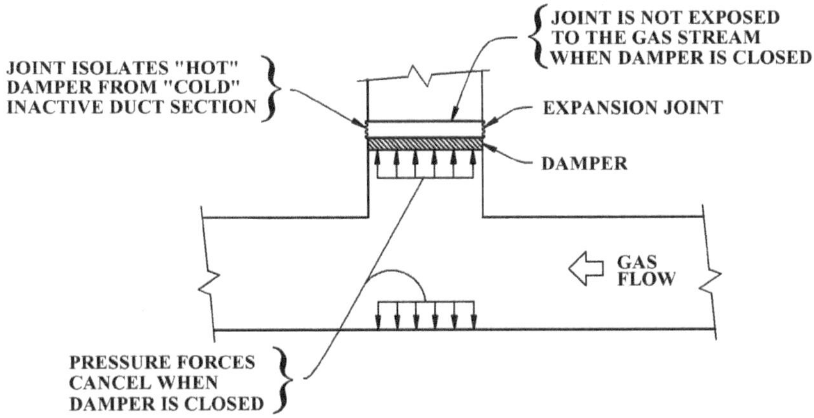

Figure 2-2. Preferred damper location relative to an expansion joint.

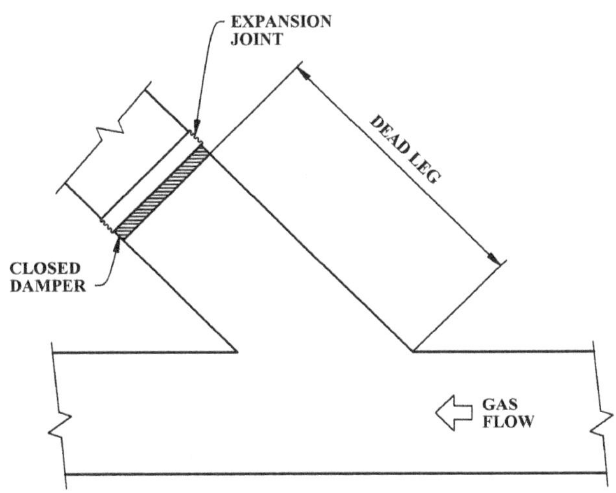

Figure 2-3. Example of a dead-leg duct section.

the flow at only one end because the damper at the other end is closed. Crossover and bypass ducts are typical locations where dead legs are encountered. By carefully locating dampers, dead legs can be avoided or at least minimized.

Some concerns associated with dead leg sections of flue-gas ducts are as follows:

- Stagnant flue-gas in the dead leg may cool off enough that its temperature is below the acid dew point, resulting in faster corrosion of the exposed internal duct surfaces.
- Cooling of stagnant flue-gas in the dead leg may introduce appreciable, undesirable thermal gradients into the ductwork structure. Thermal gradients are discussed in various sections of this book, including Section 2.7.

- Because the flue-gas flow is relatively stagnant in a dead leg, fly ash will drop out of the flow stream. The resulting accumulation will increase the gravity load and contribute to the development of thermal gradients in the duct. This effect is discussed further in Section 2.7. If possible, orient/slope the geometry of the dead leg to prevent ash accumulation. Consideration can also be given to including dampers at each end of a dead leg, with a seal air system to mitigate ash accumulation. Ash loading on closed damper blades should also be considered.
- The duct plate and internal elements in the dead leg will be exposed to the operating environment even though the duct is not conveying any flue-gas. External stiffeners will also be subjected to the operating loads and service temperatures. This exposure may shorten service life when creep or corrosion are significant factors.
- During transient pressure or excursion temperature events, the dead leg may be unnecessarily exposed to adverse conditions it might otherwise be protected from.

Dead legs in air ducts are less of a concern than in flue-gas ducts given the less severe design conditions and lack of fly ash on the air side of the combustion process.

Locating dampers in toggle duct sections should be avoided whenever possible. The three directions of movement of the toggle duct section could adversely affect the alignment of the damper frame and the damper actuator unless the actuator is mounted directly on the damper. In addition, the toggle duct section and its supports would be subjected to large forces from the damper. These forces would come from the damper weight, actuation forces, unbalanced pressure, and wind on an open guillotine damper blade or its housing. This additional loading complicates support design because toggle duct sections exhibit rotational movements and are designed to move a considerable amount, as shown in Figures 1-7 and 2-5.

2.2.5 Flue-Gas Monitoring: Instruments

Government regulations require the monitoring of flue-gas for permit compliance. In addition, the owner wants to monitor system operating data. The instruments required to collect these data can be very sensitive to stratification of the flue-gas stream. Therefore, to ensure that representative data are collected, special criteria are often followed regarding the placement of flow or emission monitoring equipment within the ductwork. For example, a common requirement is that an instrument be located in a straight section of duct a certain minimum distance from bends or significant flow obstructions both upstream and downstream of the instrument. These distances are usually tied to the duct cross-sectional dimensions, such as a multiple of the equivalent duct diameter. A length of 8 equivalent duct diameters is not uncommon, but this value should be provided by the instrument manufacturer. This criterion can dictate the placement of equipment, the arrangement of certain duct sections, or the location of internal trusses and flow distribution devices. Consequently, the flue-gas monitoring systems should be defined and conceptually designed in parallel with the development of the duct arrangements.

2.2.6 Access Doors

Most ducts have ports or doors for maintenance access. The number, locations, and types of doors depend on the duct arrangement, the type of equipment and instrumentation in the duct system, the duct maintenance needs, and the owner's preferences. A general rule of thumb is to provide a door for each horizontal level of duct run. The door locations should be coordinated with the layout of the duct support structure and the other access platforms provided for the plant. The doors and hardware are to be designed for all duct design conditions and be airtight. Doors in insulated ducts should be well insulated to prevent corrosion caused by local cold spots. Chapter 11 has more on door insulation. Hinged doors with quick-opening latches are most convenient for examination and maintenance personnel but are more difficult to seal and insulate than removable bolted panels. Davits for door panel support and chains to restrain unbolted doors from swinging also aid plant maintenance efforts. A minimum size of 24 in. × 30 in. (610 mm × 762 mm) is generally recommended, but smaller doors such as 18 in. × 24 in. (457 mm × 610 mm) are sometimes used for economy. Larger doors are sometimes used because they are more convenient for plant maintenance personnel. Handholds and ladder rungs are often welded to the duct walls, either next to or approximately 13 in. (33 mm) above the doors, to make passage easier. Doors, handholds, and ladder rungs should be located for the most ergonomic access. The door centerline should preferably be located 39 in. (99 mm) above the duct floor of a rectangular duct (Figure 2-4).

In round ducts, doors may be curved to suit the duct shape, or an extension may be used to allow a flat door. In any arrangement, the potential area for the buildup of ash behind the door should be minimized. Depending on the location of the doors, guardrails, tie-offs, or other provisions may be installed within and outside the ductwork to protect personnel from accidentally falling into equipment or down a vertical or sloped duct section. All such provisions are to be inspected prior to use, given the potential of deterioration from exposure to flue gas during plant operation. Plant and OSHA safety procedures for personnel access must always be followed.

Alternatively, struts or other fall protection barriers that function similarly to guardrails may be provided to create an obstruction to prevent plant personnel from accidently falling down a vertical or sloped section. A commonly used fall protection barrier that is normally required at a vertical or sloped section of ductwork are turning vanes. The advantage of not identifying these barriers as guardrails on design drawings or other engineering documents is that the user of the fall protection barrier must determine if it is OSHA compliant and has the necessary strength required by OSHA for fall protection (accounting for any deterioration over the plant operation). Labeling a barrier a guardrail may lead

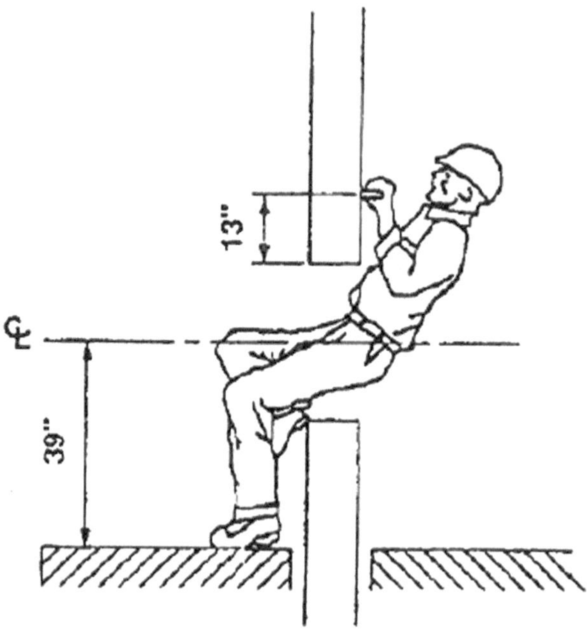

Figure 2-4. Access door location in rectangular duct.

workers to assume that the barrier meets all OSHA requirements, and thus not properly inspect it for deterioration. Technically, ductwork is unoccupied equipment, and guardrail requirements in building codes may not be applicable.

2.3 THERMAL EXPANSION

Most ducts convey heated air or flue-gas at temperatures well above ambient. Even ducts conveying combustion air are frequently exposed to elevated temperatures, because preheaters heat the air to improve combustion efficiency. Thus, the structural analysis and design of ducts at elevated temperatures must consider several significant issues to accommodate thermal expansion and contraction.

2.3.1 Expansion Joints

Expansion joints are used in a ductwork system to allow the ducts to expand and contract in a relatively unrestricted and controlled manner. Expansion joints allow thermal movements while alleviating thermally induced stresses in the equipment, ductwork, and duct support structures. They also may be used to isolate duct sections from equipment, as explained in Section 2.2.1, or to enhance the ductwork's structural determinacy, as discussed in Section 2.4.2.

Expansion joints are made of either some sort of fabric or rubber-like material, or light-gauge convoluted metal (these are termed *nonmetallic* and *metal* expansion joints, respectively). Expansion joints are preferably located so that they will not be exposed to large transverse movements. This will reduce the amount of flue-gas blockage from any internal ash shields or flow liners and reduce the shear forces inherent in metallic expansion joints.

An expansion joint is limited in the amount of movement it can accommodate. An expansion joint of a given design from a given manufacturer will be able to accommodate only a certain amount of axial, rotational, and transverse deformation before the material fails. Cold offsets can be incorporated in the duct arrangement to reduce the severity of movement to be absorbed by an expansion joint. Metal joints will typically have more severe limitations than nonmetallic joints. In addition, as discussed in Section 2.2, metal joints are much stiffer than nonmetallic joints, especially in shear, and can transmit both more load and more vibration, which should be considered in the ductwork design. The design of nonmetallic expansion joints should follow the Fluid Sealing Association's *Ducting Systems Non-Metallic Expansion Joints Technical Handbook* (2010) and manufacturer recommendations. It is extremely important to establish early in the ductwork layout stages the type of joint that will be used for each duct section and the deformation and temperature limits for each joint used. Access and insulation arrangements around the joints should also be determined.

Careful placement of expansion joints and supports gives the structural engineer control of where the unbalanced pressure loads are transmitted to the supporting structure (Figure 2-5). Locating expansion joints should also consider the ease of erection, including the location and installation sequence of the supporting structural steel and the duct. Expansion joint placement also affects the friction loads generated at the support points as bottom-supported ductwork expands and contracts. Expansion joint placement can be used to establish the level of structural determinacy of duct sections.

Expansion joints should not be added indiscriminately to a system but should be used only where they provide clear benefits. They usually are both a high capital cost item and a high maintenance item, because repair or replacement may be required every several years. They can also indirectly have a detrimental effect on the adjacent ductwork, because their need to deform makes them difficult to effectively insulate and weatherproof. This may lead to cold spots in the ductwork, which could increase thermal gradients and stresses or promote corrosion (this is discussed further in Section 11.6).

Nonmetallic expansion joints normally provide complete sealing in the bellows area but may leak slightly in the clamping areas. If more effective sealing is desired for the complete joint assembly in service conditions such as downstream of a wet flue-gas absorber with corrosive fluids entrained, then the design and installation should incorporate possible leakage testing using a foam-building liquid, with acceptance criteria agreeable to all.

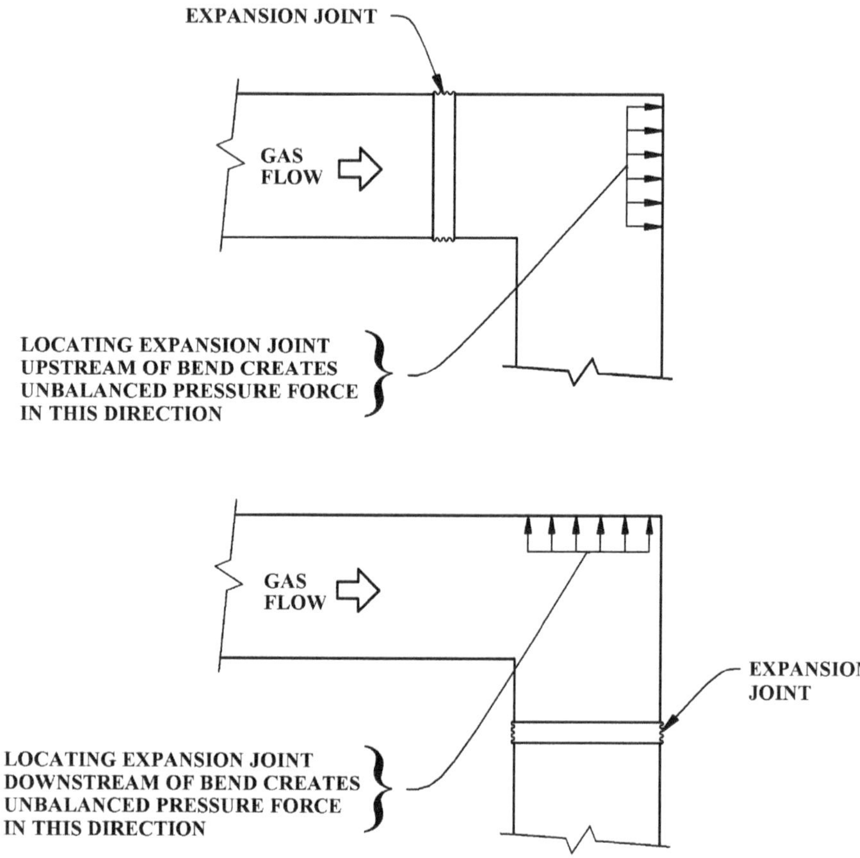

Figure 2-5. Example of the effect of expansion joint placement on loads.

2.3.2 Toggle Duct Sections

Toggle duct sections are made by installing metal or nonmetallic expansion joints on both ends of a short duct section to accommodate large differential transverse deflections between the two adjacent duct sections. The toggle duct section is usually supported from the adjacent duct sections with sliding, slotted, pinned connections, or by the shear stiffness of metal expansion joints if the loads are relatively small. The expansion joints then act like sliding hinges in the duct system. Figures 1-7 and 2-6 illustrate a typical toggle duct section.

2.3.3 Location of Anchors and Guides

The supports of a duct system generally consist of gravity supports and a combination of anchors, lateral supports, and guided supports. The gravity supports may be hangers or bottom-supported posts. The bottom supports may be the anchor, guided supports, and unguided supports. Ducts carrying air at ambient temperatures are sometimes directly attached to their support structures

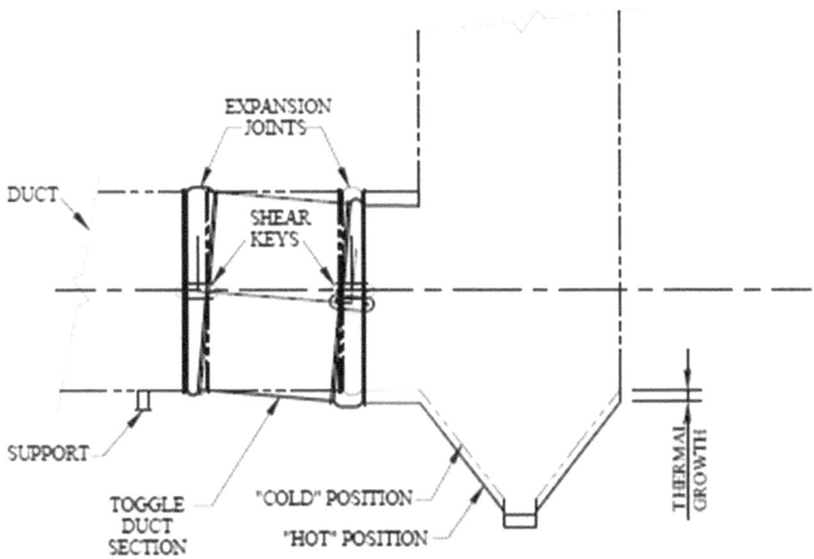

Figure 2-6. Typical arrangement with a toggle duct section.

without provisions for movement, because thermal expansion may not need to be accommodated.

All gravity loads, externally applied loads such as wind, and internally applied loads such as pressure are to be resisted while still allowing the duct relatively unrestrained thermal growth. Figure 2-7 shows a typical support arrangement for a simple, bottom-supported duct section.

Anchors and guides should form a lateral and vertical load-resisting system with the least possible restraint of thermal growth while still maintaining the structural stability of the duct section. The selection of the anchor point and the guide locations should consider the ability of the support structure and foundations to resist the duct's vertical and lateral loads; the relationship of the duct to adjacent equipment; the desired level of indeterminacy in the support system; and the effect on expansion joint deformations. The support locations need to be carefully coordinated with the support steel layout. In an efficient support structure, the duct anchors and guides are provided on the support structure's column lines. Other considerations related to duct supports are discussed in Section 2.4.

For duct sections with multiple sets of supports, it may be best to place the anchor point near the center of mass of the duct section (Figure 2-8). This will reduce the friction forces on certain duct elements.

Anchors are typically formed by bolting, welding, or otherwise securing the duct directly to its support structure. Guides are most often formed by installing bumpers or guide bars at duct sliding support points (Figure 2-9), or by loosely bolting the duct to its support structure using slotted holes. Guides and slotted holes should be oriented so that the expected thermal movement of the duct is in

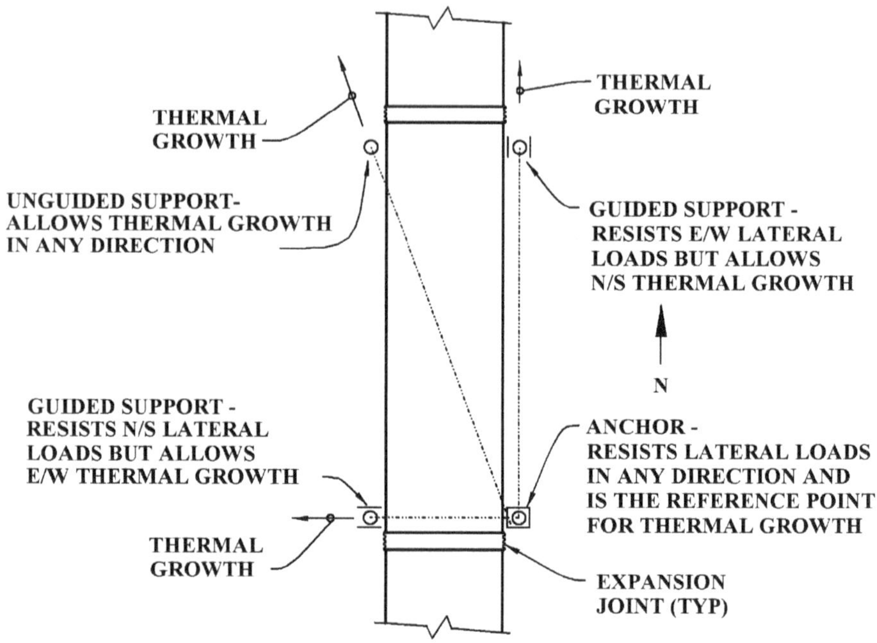

Figure 2-7. Typical duct anchor and guide arrangement.

line with the anchor point. There can be performance problems with slotted-hole details. If they are not installed properly, the bolts may shear off and the guide detail becomes ineffective. An orthogonal guide system is generally easier to fabricate and erect.

Duct guides can be arranged in such a way that a tangible anchor point is not needed. Where two lines of guides intersect, a virtual anchor location is formed even though there is no physical attachment to the support structure at this location (Figure 2-10). When guided in this manner, the duct will expand in all directions from this virtual anchor point as if a real anchor were present. The guides are to be designed for all loads considering that there is no true anchor present.

2.3.4 Slide Bearing Plates

Slide bearing plates are commonly used to reduce the frictional forces that develop between the base plates of bottom-supported ducts and the support structure as the duct expands and contracts under temperature changes. Many types of slide bearing plates are available. Most have a maximum coefficient of friction between 0.05 and 0.15. The actual coefficient of friction varies with many factors, including the mating surface materials, the age and condition of the slide bearing plate, the service temperature, and the bearing pressure. The coefficient of friction that is assumed in the structural analysis should be based on the range of values expected over the plate's entire design life, which may not be the same as the value published for a new plate.

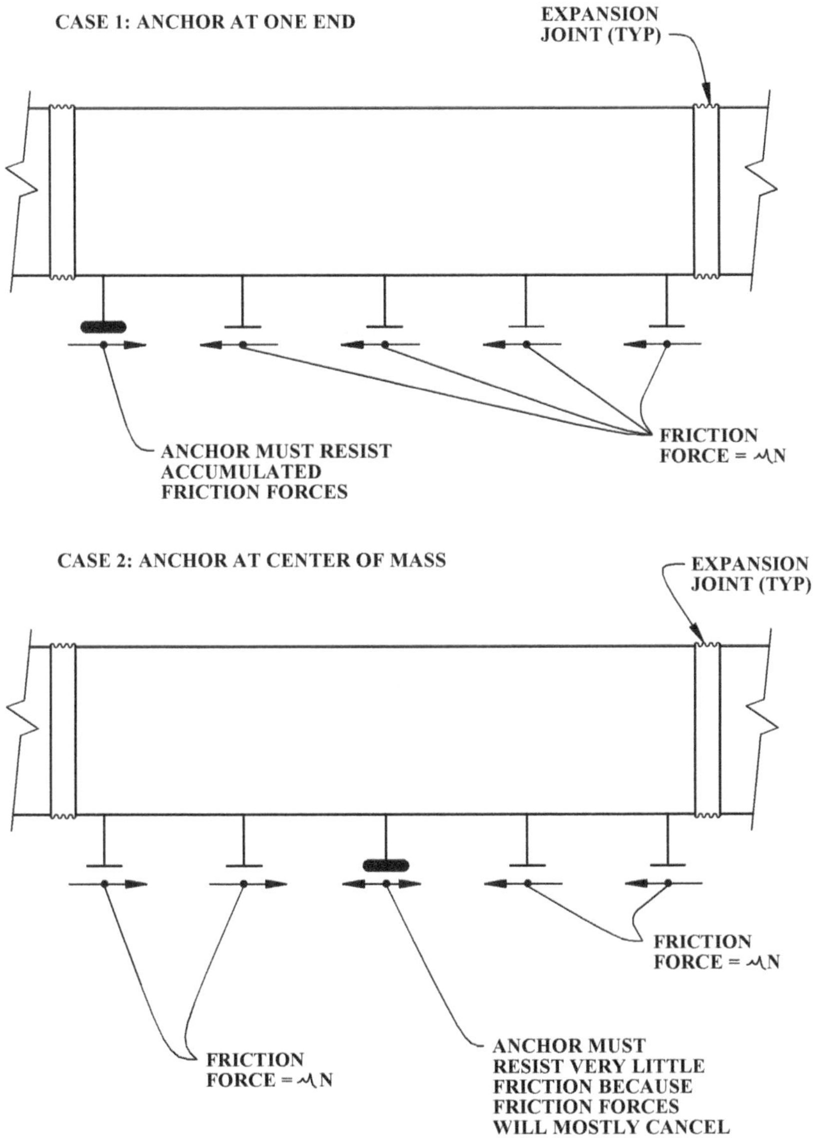

Figure 2-8. Locating the anchor point at the center of mass.

Two shapes of slide bearing plates are available: flat and spherical. Flat slide bearing plates are used when little or no relative rotation is expected at the mating surface. If appreciable rotation of one surface relative to the other is anticipated, then spherical slide bearing assemblies may be specified. A bearing assembly may consist of two separate sliding surfaces (Figure 2-11). One is a flat surface that allows relatively free translation, and the other is a spherical surface that allows relatively free rotation.

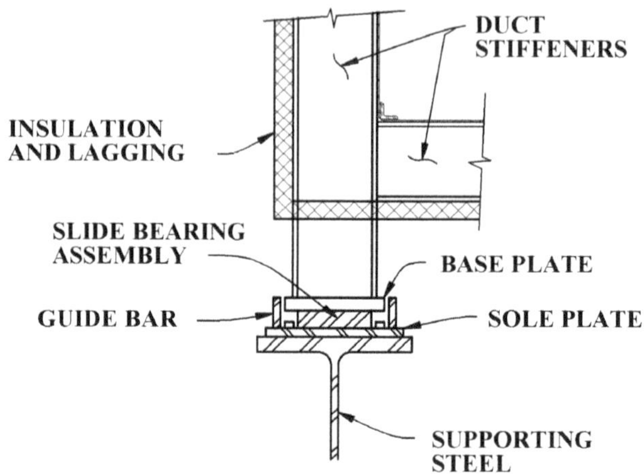

Figure 2-9. Typical guided bottom support detail.

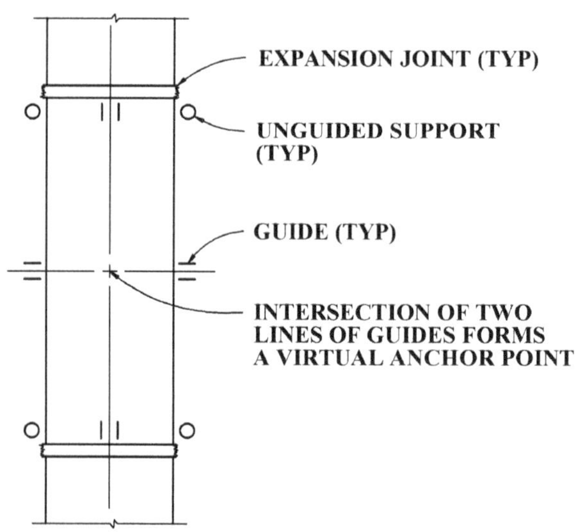

Figure 2-10. Virtual anchor point.

Slide bearing surfaces are typically either polytetraflouroethylene (PTFE or Teflon) or self-lubricated. PTFE slide bearings are typically more cost advantageous but are not recommended when the temperature at the slide bearing surface exceeds 500 °F (260 °C).

Slide bearings can be procured with built-in lateral or uplift restraints for certain loading and movement configurations. This can simplify the support assembly design and arrangement but may be more costly than a built-up support assembly with an unrestrained slide bearing.

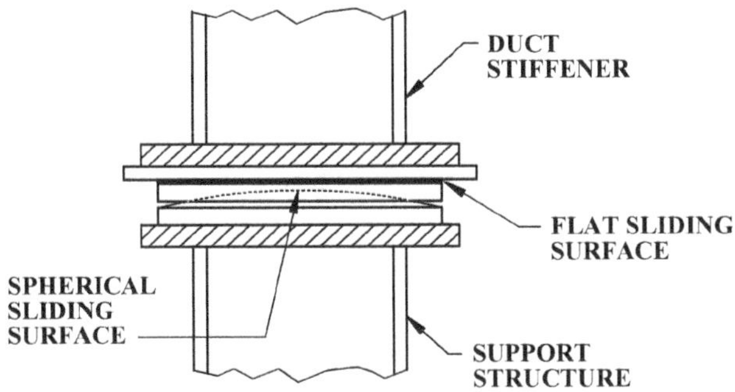

Figure 2-11. Typical spherical and flat slide bearing plate assembly.

Roller duct supports are sometimes used in lieu of slide bearing plate assemblies. The performance of rollers is sensitive to their orientation, because relatively unrestricted thermal movement must occur for the support to function properly as designed. The rigidity and hardness of the roller are important to its design, because a roller that will deform under the duct's loads may not operate freely. Roller design details should include provisions for keeping debris out of the roller's path. With the current successful performance of relatively inexpensive slide bearing plate assemblies, rollers are seldom used today under ducts.

2.3.5 Clearances

When arranging a ductwork system, the design team is to consider that ducts conveying air or flue-gas at elevated temperatures are not stationary structures. Because of thermal expansion, some points on the ducts may move several inches in all three directions between their cold and hot positions. The expected movements are to be calculated, and adequate clearance to adjacent structures and equipment provided. In addition to considering thermal expansion and contraction, the calculation of clearances needs to include adequate margins for final stiffener sizes, insulation and lagging details, and fabrication and erection tolerances. Structural deflections of the ducts and their support structure may also need to be accommodated if they are expected to be significant. Usually, the structural deflection of ductwork is negligible, but often the duct's supporting structure has some flexibility. Adequate clearance for construction access to erect the ducts and install the insulation and lagging also needs to be provided. Adequate clearance is also needed at duct access doors, for access platforms, and for equipment maintenance access.

2.3.6 Support Eccentricities

As discussed, ducts that convey air or flue-gas at elevated temperatures will experience thermal expansion on unit start-up. This three-dimensional expansion will result in a change in the position of all duct supports that are not anchored. At the same time, however, the support structure is at ambient temperature and thus remains stationary. As a result, the eccentricity of the duct support with respect to its support structure may change as the duct heats and cools. Significant stresses may result in both the duct and the support structure because of the moments and torsional couples induced by this eccentricity. The structural engineer needs to be aware of this phenomenon and account for it in the design of both the duct support elements and the duct support structure. A common practice to minimize the effects of eccentricities induced by thermal expansion is to initially offset the duct support point from its supporting structure by an amount equal to one-half of the expected movement at the duct's operating temperature. In this way the eccentricities in the hot and cold conditions are roughly equal, but opposite. This arrangement is shown in Figure 2-12. In this same figure, the effect of the load eccentricity on the support steel could be reduced by centering the lower slide bearing plate on the support steel member and ensuring that the duct support leg is sufficiently rigid, limiting its angular deflection.

2.3.7 Attachment of Cold Items to Ducts

A great deal of care is to be taken when items such as electrical conduits, ladders, and access platforms are supported from ducts conveying air or flue-gas at elevated temperatures. Details should be used that allow differential thermal expansion and contraction between these items and the hotter ducts. Otherwise, significant forces may develop, and failures may result. Elements that remain

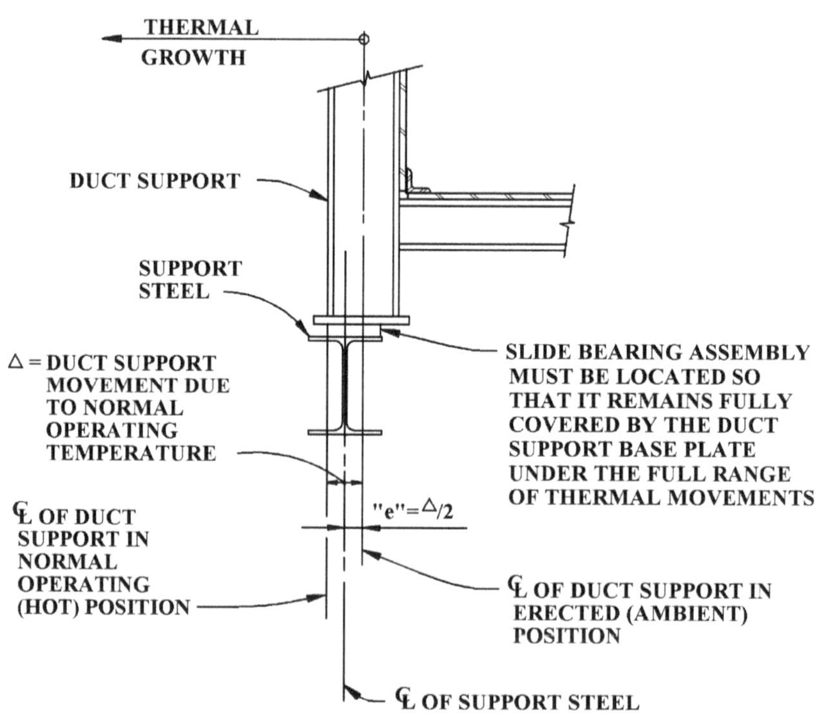

Figure 2-12. Initial offset of bottom support to reduce eccentricity.

essentially at ambient temperatures should not be rigidly attached at two or more locations to ducts conveying air or flue-gas at elevated temperatures. Typically, one end will be rigidly attached to the duct, while the other end allows the duct to move freely in the direction of thermal expansion.

Similarly, duct guides or bumpers, support brackets, and duct support legs, which extend outside a duct's insulation and lagging will be exposed to the ambient air and thus may be at a lower temperature than the duct to which they are attached. Construction and insulation details for these elements are to be designed considering this potential temperature difference, or else damage to these structures could result.

2.4 SUPPORTS

Most ductwork is supported by structural steel framing, although occasionally an arrangement will allow the ducts to be supported directly on the foundations. The duct support structure layout should be developed in conjunction with the ductwork arrangement to optimize support locations, support types, foundations, and access platforms.

The design of structures supporting ductwork needs to consider the following factors, which are usually not significant in the design of typical building systems.

- Torsional effects are almost always present, in at least some parts of the structure, because of the frictional forces and support eccentricities that may develop as the ducts expand and contract.
- Localized stresses, such as local flange or web bending, may be significant at duct support points, because of the large, concentrated forces and support eccentricities.
- Heat transfer from the ducts to the support structure can result in high thermal stresses in the support steel. This effect depends on the operating temperature of the duct, the geometry of the interface, and the duct insulation details. It is often encountered with support members framing through the crotch of hoppers in hot ducts.
- The stiffness of structures and foundations supporting indeterminate ducts can have a profound effect on the distribution of loads to the duct support structure. This includes arrangements where multiple elevations of duct supports or guides are employed for an individual duct section.
- Duct support structures may have unique temporary loads because of duct construction methods or sequencing.

The following sections discuss some of the other significant issues the structural engineer needs to address when determining how to best support the ductwork.

2.4.1 Support Locations

For the most efficient duct support structure, the duct supports should be located on support structure column lines whenever possible. This will create the most direct load paths to carry the duct loads to the foundations. The duct support locations and support structure framing should also be arranged to allow the friction and unbalanced pressure forces to cancel as directly as possible (Figure 2-13). This will minimize the areas of the support structure's bracing and foundations that need to be designed for these loads.

Ducts often bridge between two independent structures that have varying drift vectors. If possible, the supports for each section of duct from expansion joint to expansion joint should be supported from the same structure to simplify support design.

2.4.2 Structural Determinacy of Ductwork

Structures are *statically determinate* when the accuracy of their structural analysis is independent of the stiffness of the structure itself and its supports. In reality, because ducts are three-dimensional stiffened plate structures, few are truly statically determinate. However, as discussed in Chapter 6, structural engineers commonly analyze ductwork as a system of planar structures, which can each be

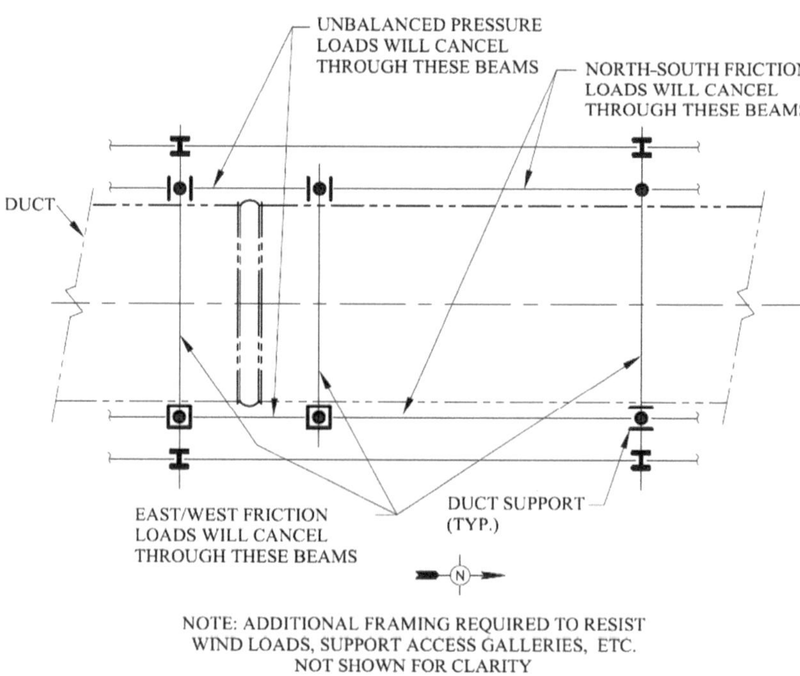

Figure 2-13. Example of support steel framing that allows friction and unbalanced pressure forces to cancel.

individually classified as determinate or indeterminate. This approach is similar to the common practice of taking a three-dimensional building structure and analyzing it as a series of determinate horizontal floors and vertical frames. One of the most significant decisions to be made by the structural engineer is whether the ductwork will use determinate or indeterminate support systems.

Indeterminate Ducts. Indeterminate structural systems often have a lower initial cost than determinate systems, because there is usually less material. If arbitrary minimum stiffener sizes or plate thicknesses do not apply, an indeterminate duct will usually be lighter, and therefore less costly, than a similar determinate duct. However, the behavior of indeterminate ducts can be unpredictable, especially in hot flue-gas ductwork applications. This unpredictability can lead to problems with locally overstressed components, potential failures, and increased maintenance. For these reasons, structurally determinate ductwork systems should usually be chosen whenever practical. When the use of indeterminate duct sections is unavoidable, or when indeterminate support systems are chosen for their apparent economy, the structural engineer should attempt to perform the structural analysis of the ducts and support structures as accurately as possible. This effort may result in a more rigorous analysis than a determinate system would require, because the validity of the assumptions on which an indeterminate structural analysis is based will have a significant impact on the structural behavior. This approach is discussed in more detail in Chapter 6.

When indeterminate ductwork systems are used, factors such as the duct stiffness, the support structure stiffness, foundation settlement, and thermal gradients within the ductwork all have a significant impact on the distribution of forces through the structural system. These unpredictable situations often lead the structural engineer to build extra contingencies into the design of the support structure and the ducts that may negate the intended material savings of using an indeterminate system. For example, a duct with multiple sets of supports may be designed for the full positive moment from the global analysis, $WL^2/8$, ignoring that the duct's continuity could allow a design moment as small as $WL^2/12$. Another typical example is where the support loads may be increased by some factor to account for any potential modeling inaccuracies or variables in the ductwork structural analysis. These contingencies would be a way to allow for the possibility that the actual distribution of the loads to the supports may vary from that calculated in the structural analysis. The application of such contingencies would tend to reduce the indeterminate duct system's cost advantage over a determinate duct system.

A straight duct section supported at three locations (six points) along its length, as depicted in Figure 2-14, should be considered structurally indeterminate. To accurately analyze this duct, the structural engineer calculates or assumes the stiffness of both the duct and its support structure. This can be a time-consuming, iterative process. In addition, especially for foundations, the stiffness values may not be reliable. If the actual stiffnesses in the as-built structure do not match those used in the analysis, the analysis results may be invalid. Similarly, if the foundations under the supports do not settle in the same way that was

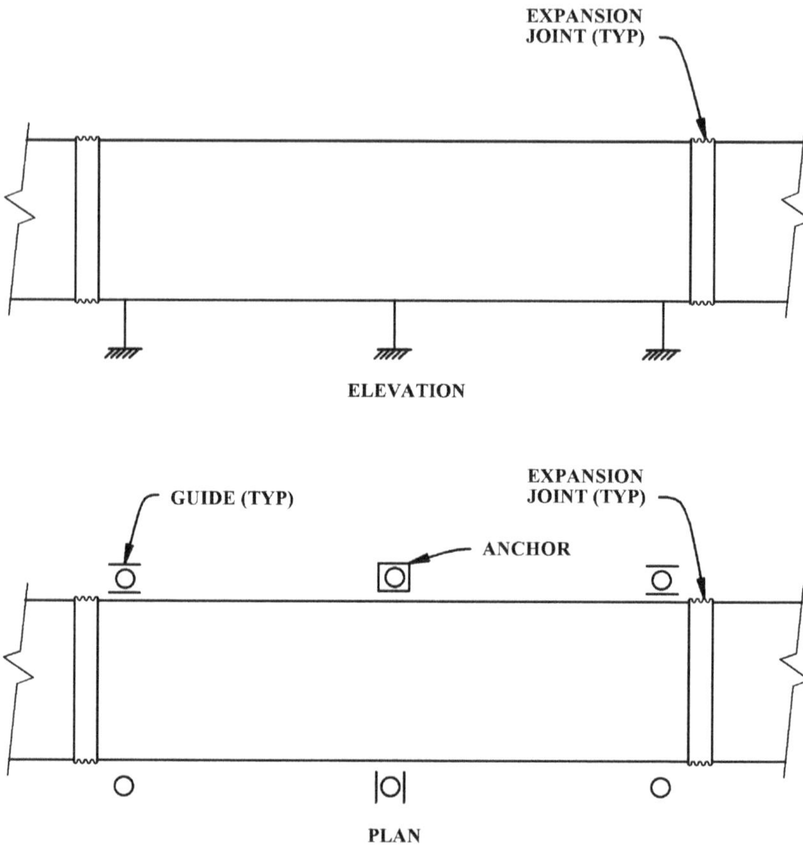

Figure 2-14. Example of an indeterminate duct section.

assumed in the analysis, the distribution of loads to each support point will change. The accuracy of settlement calculations depends on the structural engineer's ability to accurately estimate the magnitude and duration of loads, but some of these loads, such as live load, ash weight, and wind, may actually be extremely variable. If the duct is subjected to a thermal gradient, the gradient will try to bow the duct. This bowing will then cause the gravity loads to redistribute among the various supports. The construction sequence or method may also affect the dead load distribution in an indeterminate duct section, because most indeterminate ducts are erected in several pieces that are then field spliced in place after they are placed on the supports. Nonadherence to fabrication and erection tolerances may also affect load distribution. Because all of the above factors can be present simultaneously, the inaccuracies introduced by using a structurally indeterminate duct system can become significant.

Determinate Ducts. If the same duct is supported at only two locations (four points) along its length, as shown in Figure 2-15, it can usually be considered structurally determinate. The distribution of loads to each support would be

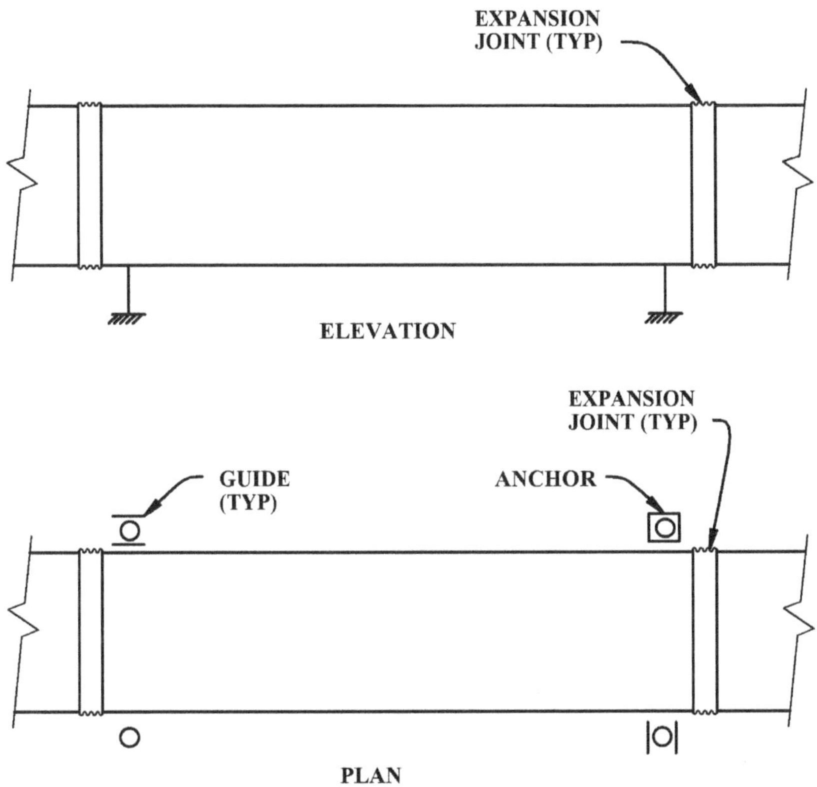

Figure 2-15. Example of a determinate duct section.

relatively unaffected by duct stiffness, support stiffness, foundation settlement, thermal gradients, or construction sequencing. Although the same size duct would have to span farther than it would if it had multiple support points, the analysis results would be more reliable. Behavior of the duct section would be more predictable. Contingencies added to the design could be lessened, and the potential for overloading components would be reduced.

To summarize, although indeterminate ductwork has been used successfully in many installations, statically determinate ducts exhibit more predictable behavior and are therefore less likely to experience local overloading and potential failures. For this reason, determinate systems are encouraged whenever practical. Where indeterminate ducts are unavoidable, the structural engineer can perform a structural analysis and design that takes all potential concerns into account. To compensate for varying calculated structural steel deflections at the different support points in both determinate and indeterminate duct support systems, the installation of shims at selected support legs may be prescribed by the structural engineer.

2.4.3 Hangers versus Bottom Supports

Ducts may be supported in a variety of ways. The best way to support a particular duct section depends on many project-specific factors. However, most duct sections can be classified as either hanger supported or bottom supported. Regardless of which support type is employed, anchors and guides are needed to control thermal expansion and resist lateral loads.

Hangers. There are three basic types of hangers: rigid, variable-support spring, and constant-support spring. Hangers are used in indoor applications more often than in outdoor applications because support steel above the ducts is usually more available inside buildings.

Hangers have the following advantages over bottom supports.

- Hangers are efficient structural members, because they are loaded in pure tension.
- Using hangers near the top of vertical sections of ductwork may result in a more economical duct design, because the duct plate will be in tension under gravity loads.
- Hangers can easily be designed to allow the hanger to pivot about its ends by using pinned connections or rocker details. This is a simple, effective way to provide a true pinned connection and accommodate thermal expansion of the duct.
- Frictional forces are practically eliminated with hangers. Pin rotational friction is still present, but it is usually small and is typically neglected.
- The use of constant-support spring hangers can help mitigate the uncertainties associated with indeterminate duct support systems.
- Hangers are usually more cost-effective than saddles for supporting circular ducts.
- Hung ductwork can usually absorb energy during seismic loading better than bottom-supported ductwork, because there is typically a significant amount of damping in the pendulum effect as the ducts sway on their hangers. Also, it will not be catastrophic if a lateral bumper fails, because the hangers will continue to keep the duct from permanently displacing laterally.

The following disadvantages are associated with using hangers as duct supports.

- Failures associated with hangers are likely to be severe, because tension failures are often a total fracture of an element. Once a hanger or gusset plate fractures, its total load must be redistributed to the other supports. If a progressive failure is triggered, there is often nothing to prevent the duct from dropping a significant distance and damaging other areas of the plant.
- The duct support structure receives its loads at a higher elevation, which may penalize the support structure and foundation design.

- Spring hangers require periodic adjustments and maintenance. They also require careful installation.
- Hangers alone usually cannot resist horizontal loads, so bumpers or lateral ties may need to be added.
- Hung ducts can be more difficult to erect than bottom-supported ducts, because the supporting steel must be erected before the ducts can be hung from it. This could create access restrictions that would preclude the preassembly of large duct sections.
- Hangers do not resist uplift unless struts designed for the compression load are used.
- Spring hangers may not resist variable ash and pressure loads, as these types of loads will be carried by the stiffer supports in the duct restraint system.

When hangers rotate with thermal expansion of the duct, they need to be long enough to keep the angle of rotation relatively small. The industry norm is that hanger rotation should be kept to less than about 5° to minimize the lateral force component of the load. Also, as described in Section 2.3 for bottom supports, hangers may be installed with an offset to minimize the effects of rotation. If a hanger is allowed to rotate excessively, the angle will induce large lateral forces into the support system, and the duct will also try to lift up (Figure 2-16).

The structural engineer should ensure that there is enough clearance in the hanger pin holes if the details are oriented such that the thermal movement of the duct is not in the same direction as the clevises would normally rotate about their support pins.

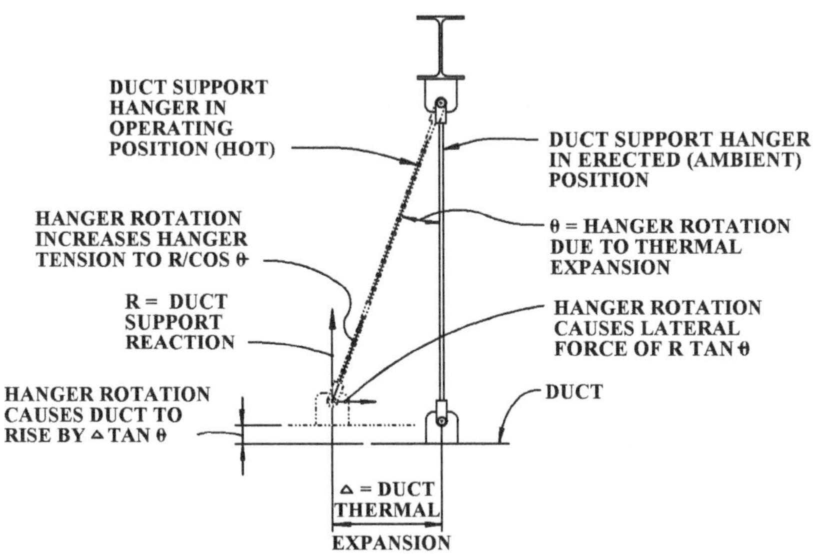

Figure 2-16. Effect of hanger rotation.

Bottom Supports. Bottom supports are generally achieved by extending selected duct side wall stiffeners past the bottom of the duct. Sometimes an independent post comes off the stiffened duct bottom. A base plate is attached that bears on the support structure, usually through a slide bearing plate assembly (this detail is shown in Figure 2-9). A system of anchors and guides is used to control thermal expansion, and slide bearing plates are usually used to reduce the frictional forces at the base plates. This type of support system is discussed in Section 2.3 and shown in Figure 2-7.

Bottom-supported ducts have the following advantages over hangers.

- Failures associated with the bottom supports are unlikely to result in total loss of support for the ductwork.
- Bottom-supported ducts are relatively easy to erect, because large sections can be preassembled and set directly on top of the support structure.
- The duct support structure receives its loads at a lower elevation, which may result in a more economical support structure and foundation design.
- Anchors and guides can be easily incorporated into the bottom support base plate designs without having to add bumpers or lateral ties.

The following disadvantages are associated with using bottom supports for ducts.

- Frictional forces will develop at the support points as the ducts expand and contract thermally. Even with low-friction slide bearing plates, these friction forces can be sizeable.
- Bottom supports often load the support structure eccentrically, because the supports move as the duct thermally expands, while the duct support structure remains stationary. This can be mitigated by careful placement of bearing plates at the interface design.
- Actual load distribution uncertainty is increased in both determinate and indeterminate duct support systems by the fixed support leg length and the lack of support height adjustability. This uncertainty can be reduced by the use of shims at the support locations.
- Vertical offsets encountered during construction are usually more difficult to accommodate in a duct section when bottom supports are used.

2.4.4 Spring Hangers versus Rigid Hangers

Rigid hangers are usually preferred, because they require little or no maintenance and they cost less than spring hangers. However, spring hangers may be appropriate under certain circumstances. The most common reason to use spring hangers is to accommodate vertical thermal movements at the support points. Constant-support spring hangers may be used to alleviate the uncertainties associated with the analysis of indeterminate ductwork systems.

2.4.5 Support Elevations

Whenever possible, all vertical supports for sections of ductwork that convey air or flue-gas at elevated temperatures should be located at a common elevation on the duct. This is the best way to accommodate thermal expansion and contraction. As a minimum, all vertical supports should be in the same plane. If they are not, then one or more of the supports will try to lift off the support structure as the duct thermally expands. This lack of support contact can result in significant load redistributions.

Having all supports at the same elevation is especially important for bottom-supported ducts, because slide bearing plates work best when the mating surfaces remain parallel. Relative rotation between the mating surfaces will result in uneven loading on the slide plate, which may damage the slide plate or cause it to lock up. For ducts with bottom supports at different elevations, this condition can be alleviated by installing the slide plates at an angle in line with the anchor point (as shown in Figure 2-17) or by installing combined spherical and flat slide bearing plate assemblies instead of just flat slide bearing plates. However, the structural

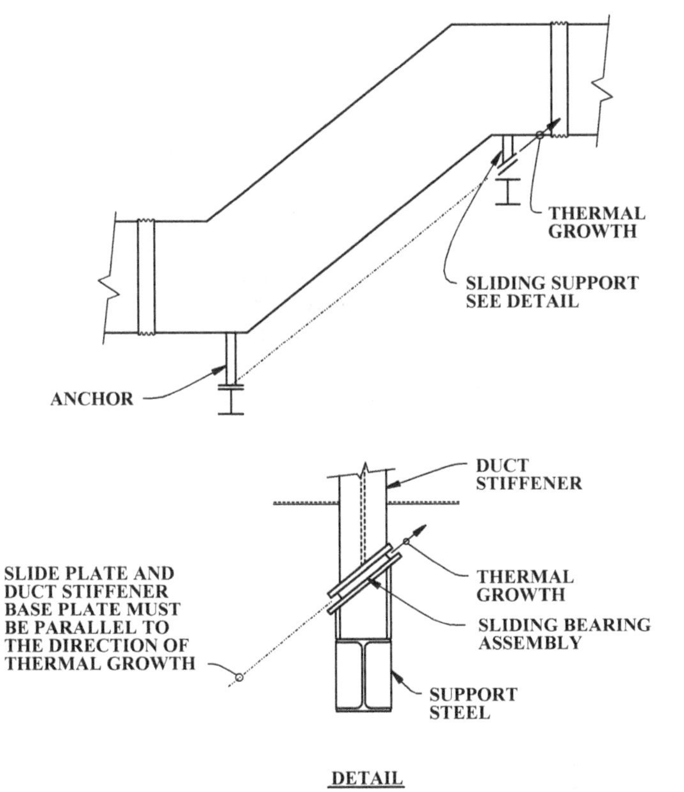

Figure 2-17. Example of the proper arrangement of supports at different elevations.

engineer should consider that fabricating and installing base plate details at precise angles is difficult; slide plates installed at an angle may induce large, undesirable lateral gravity forces into the support elements; and spherical bearing assemblies cost more than flat slide plates. Using hangers instead of bottom supports could alleviate these conditions.

2.4.6 Support Structure Stiffness

The stiffness of the support structure is usually more of an issue if the duct is structurally indeterminate, as discussed in Section 2.4, or if the supports of a duct section have vastly different stiffness, as shown in Figure 7-2. For most ducts, the support structure can be designed for standard building deflection criteria. When support structure stiffness is important, the stiffness of the foundations and their settlement should also be considered. Support structure stiffness may also need to be considered if the support structure is relatively flexible. For example, if a tall, slender tower is used to support a duct section, the lateral deflection of the tower under wind, seismic, or unbalanced pressure loads may need to be factored into the expansion joint design or into the calculation of clearances to adjacent structures.

2.4.7 Ductwork Stiffness

Given their size and material thickness, ducts are by nature very stiff structures. The overall behavior of ducts is similar to that of a large box girder. Overall stiffness or deflection checks of air and flue-gas ducts are generally not required, although local checks of the plate and structural elements are sometimes necessary to avoid vibration problems. Lateral support of a duct section at more than one elevation may be necessary. However, the structural engineer should consider the relative stiffness between the duct and the structure and the effect on load transfer. The calculation of overall deflections at expansion joints, dampers, equipment flanges, or other locations where excess movement could cause problems may be prudent.

Some structural engineers choose to apply indirect stiffness criteria to the design, such as requiring that the distance between duct supports be no more than some multiple of the duct height or width.

2.4.8 Supporting Ducts from Equipment

Ordinarily, ducts should be supported independently from equipment to simplify the interfaces, reduce the transmission of vibrations, and avoid the application of loads to the equipment that could affect performance. However, there are some instances where it makes sense to use equipment to support certain ductwork sections. For example, the transition duct sections at the inlet and outlet of a precipitator are commonly supplied with and supported by the precipitator. Whenever equipment is used to support ductwork, the ductwork arrangement and loading information should be communicated to the equipment supplier to ensure proper integration into the equipment design and to avoid equipment warranty problems.

2.4.9 Retrofitting Ductwork

While developing the ductwork arrangement for a retrofit project, there may be several constraints created by the arrangement or by the capacity of the existing structures. The location of existing steel framing and the capability of existing foundations may dictate the location for the new duct supports, because the ability to erect new support framing or install new foundations may be severely limited. Also, crane access for the duct erection may be so limited that the construction methods become a significant factor in the ductwork arrangement and design approach. See Chapter 13 and ASCE Retrofit Task Committee (1991), which provide excellent guidance in this area.

2.5 DUCT GEOMETRIES

Ductwork is generally either circular or rectangular in cross section. The exceptions to this are transition configurations. The selection of a basic cross section should be made on the basis of its effect on flow, cost, and structural behavior, while also considering external constraints such as available space, available means of support, connection to equipment, and fabrication, shipping, and erection limitations. This section discusses some of the ways the duct's shape influences its behavior or cost. The process engineer should be consulted for flow considerations.

2.5.1 Basic Ductwork Configuration

The fundamental decision regarding a duct's shape is whether it should be circular or rectangular in cross section. Circular ducts with minimal stiffeners often have the lowest total installed cost for smaller-diameter ducts, which can be shop fabricated and shipped in preassembled sections. For larger ducts, fabrication, shipping, or erection limitations may make a circular duct less economical. Also, space constraints in the area through which the duct must be routed will often preclude the use of round ducts. Shipping limits are addressed in more detail in Chapter 10. Where permitted by the arrangement, a square cross section is generally preferred for rectangular ducts, because it provides the most efficient section for both flow and structural considerations. Square ducts also sometimes offer economies related to repetitive detailing and fabrication for all four sides.

Following are some of the significant items that affect the overall cost of the ductwork and should be reviewed when making these basic decisions:

- Effect on the plate thickness of various stiffener spacings and sizes—perhaps even a design with no stiffeners;
- Amount of welding required for a given stiffener arrangement;
- Transportation mode, the extent of temporary bracing required for shipping and erection, and the maximum shipping piece size;

- Methods available for supporting the ductwork, either below or above the duct;
- Effect on load magnitudes for ash, wind, and snow, among others, that the duct and support structure will need to support; and
- Relative cost of shop labor versus field labor, including the fabrication and erection of the duct and its support structure, and the application of insulation and lagging.

2.5.2 Rectangular Ducts

Ducts with rectangular cross sections (Figure 2-18) are commonly used in utility and large industrial boiler applications. Duct panels are typically fabricated from flat plate, with rolled steel shapes welded to the exterior or interior at periodic intervals to stiffen and support the plate. The corners are often reinforced with angles or bent plates, and internal trusses may be used to maintain the squareness of the duct, transmit pressure forces, and transfer loads to the support structure. The completed duct sections behave like large box girders, so they have excellent strength and stiffness. For more information on the behavior, analysis, and design of rectangular ducts, refer to Chapters 6 and 7.

Rounded corners are often suggested to reduce pressure drop at bends in the ductwork. Typically, the mechanical process engineer might want both the inside and outside corners at duct elbows to be rounded. However, rounded corners are more difficult to analyze, design, fabricate, and erect than square corners. Therefore, their use should be limited to areas where flow-related benefits are significant and justifiable.

2.5.3 Circular Ducts

Circular ducts are most often used when the outside diameter is 12 ft (3.6 m) or less, although circular ducts with larger diameters are sometimes used.

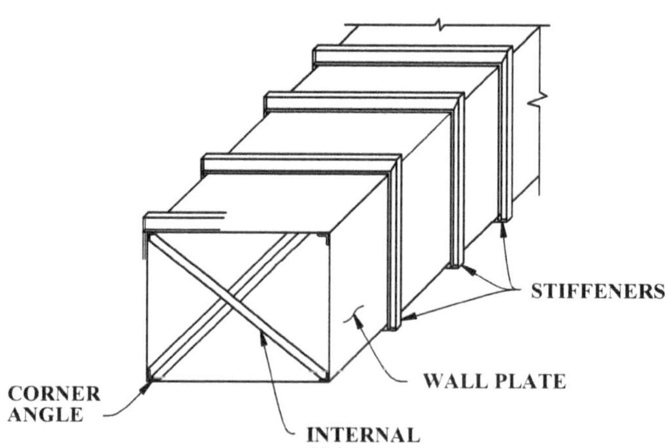

Figure 2-18. Typical rectangular duct construction.

The maximum diameter for a cost-effective design is strongly influenced by space, fabrication, shipping, and erection limitations. This is discussed in more detail in Chapter 10. Here are some of the advantages of using circular ductwork.

- Circular ducts are efficient in resisting positive internal pressure loads. This efficiency may allow the use of a thinner wall plate than would be needed for a rectangular duct.
- Circular ducts usually require fewer and smaller stiffeners than rectangular ducts. Depending on the loading, duct diameter, and wall thickness, stiffeners may even be eliminated altogether. Buckling loads under negative-pressure design conditions is also to be considered in the design of the shell and stiffener spacing and sizes.
- For a given cross-sectional area, a circular duct will have a smaller perimeter than a rectangular duct, which may reduce the duct weight and the insulation and lagging cost. The lower weight should result in a less costly duct; it may also reduce the support structure requirements.
- Circular ducts usually have better internal flow characteristics than rectangular ducts, and thus fewer turning vanes and smaller pressure drop.
- Better shape factor of a circular duct should result in smaller wind loads than a rectangular duct with an equivalent projected area, although the projected area of a circular duct may be larger than that of a rectangular duct to obtain the same internal flow area. This lower wind load may allow a more economical overall duct or support structure design.
- Circular ducts will not retain as much snow as rectangular ducts of the same width, which may allow a smaller design snow load. Again, this may result in a slightly more economical duct design and a lighter duct support structure.
- Rainwater and snowmelt will drain off a circular duct much better than off a flat or slightly sloped rectangular duct. This means that there is generally less chance of water-related damage to the insulation and to the duct plate with circular ducts.

There are also disadvantages to consider:

- Transitions from a circular shape to a rectangular shape or to a rectangular equipment flange can often be difficult to design, fabricate, and erect.
- To maintain the same cross-sectional area for flow, a circular duct must be approximately 13% wider internally than a square duct. This wider duct could cause layout constraints and erection problems.
- Designing access doors and providing access to instrument connections may be more difficult for circular ducts.

Stiffeners for circular ducts are typically fabricated from plate or relatively small rolled shapes. More information on the cost and structural design of circular ducts is given in Chapter 7.

2.5.4 Hoppers

Hoppers are often inserted into the duct system where ash is expected to drop out of the flow. Hoppers are usually either conical or pyramidal and may have ash removal equipment suspended from their outlets. Hoppers may also be fitted with rappers, air cannons, or other devices to help the collected material flow better during removal.

2.5.5 Transition Duct Sections

Unusually shaped duct sections are often required to provide a relatively gradual transition between ductwork and equipment or between two duct sections with different cross sections. Abrupt transitions should be avoided because of the potential adverse effects on flow, erosion, ash accumulation, and pressure drop. For a discussion of ductwork transitions at large fans, refer to Section 2.2.

2.6 INTERNAL TRUSSES AND STRUTS

Ductwork internal trusses and struts are used to transfer loads between various components of the duct. They are also used to maintain the squareness or roundness of the duct. This section addresses some of the issues to be considered when locating and arranging internal trusses and struts. The structural design of truss members and struts is covered in Chapter 8. Other internal elements, such as turning vanes, perforated plates, and other flow distribution devices, are typically laid out by the mechanical process engineer, often with the aid of flow model tests. However, the structural engineer should have input on their locations, especially when they are to be located in areas near internal trusses and struts. The structural design of flow distribution devices is discussed in Chapter 9.

2.6.1 Location and Layout

Truss diagonal members and struts are typically fabricated from pipe sections to minimize disruption of the air or flue-gas flow. Sometimes angle shapes are used, with the heel of the angle facing upstream. This orientation provides a better shape factor for flow resistance but also subjects the angle to much higher abrasive wear than using a flat plate oriented normal to the flue-gas flow. The arrangement of internal trusses and struts should be closely coordinated with the arrangement of flow distribution devices to minimize interference and understand their effect on flow.

The structural engineer should always attempt to minimize disruption of the flow when laying out internal trusses and struts. Members and their connections should present the smallest projected area as practicable.

Trusses are typically, but not always, used in the following locations.

- At each pair of duct supports, to maintain squareness and transfer the reactions from the duct to the supports;

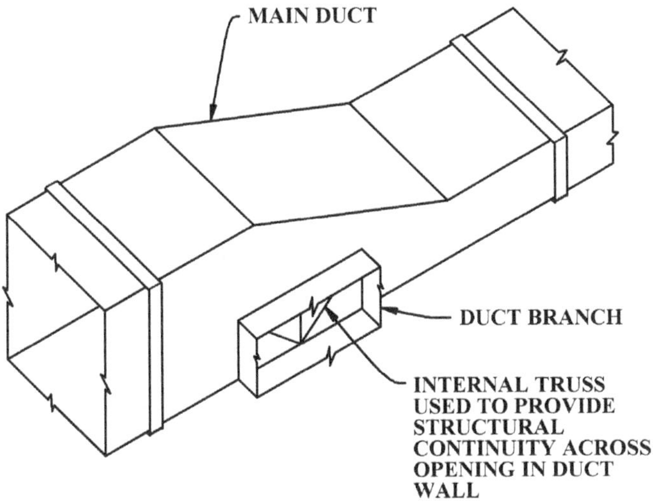

Figure 2-19. Internal truss at a duct branch.

- At other locations where maintaining squareness is important, such as at connections to equipment or expansion joints—however, as discussed in Sections 2.2 and 2.7, trusses should normally be avoided at major fan inlets and outlets;
- At branches of a ductwork system where the opening for the adjoining duct creates a structural discontinuity in the duct wall (Figure 2-19);
- Along the centerline of a very large duct section, to reduce the span of the stiffeners and help transfer the loads to the supports—this arrangement may be valuable when there are restrictions on stiffener depths because of limited clearance around the duct; and
- Inside ducts with complex geometry because of limited space or having to retrofit new ducts into existing plants—in these cases, trusses are often helpful to transfer loads around corners and through transitions.

Internal struts are often used to cancel pressure loads applied to the two opposite sides of a duct, to reduce the span of the duct stiffeners, or to allow stiffeners on opposite sides of a duct to share in resisting loads applied to only one side (Figure 2-20). Used in this manner, the cost of the struts should be weighed against the corresponding reduction in external stiffener sizes and additional pressure drop. Internal struts are also commonly used in circular ducts at periodic intervals or at support locations to prevent ovaling. As with trusses, internal struts should be avoided at the inlets and outlets of major fans.

Truss arrangements should also provide some flexibility under potential thermal gradient conditions. Some common examples of flexible and inflexible truss arrangements are shown in Figure 2-21. Work points should be located to achieve symmetry or at points of concentrated loads. Work points at the duct walls

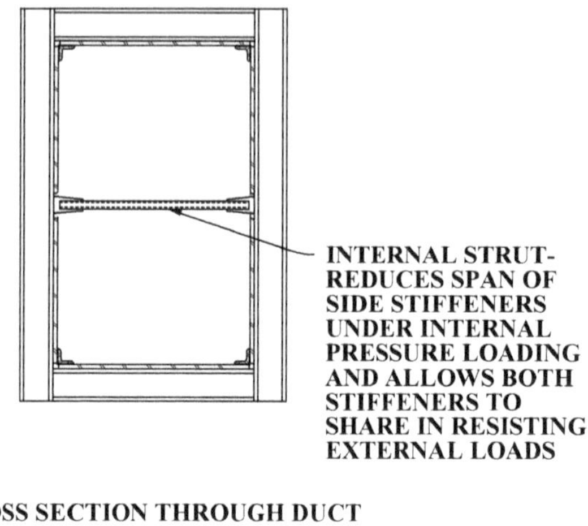

Figure 2-20. Internal strut.

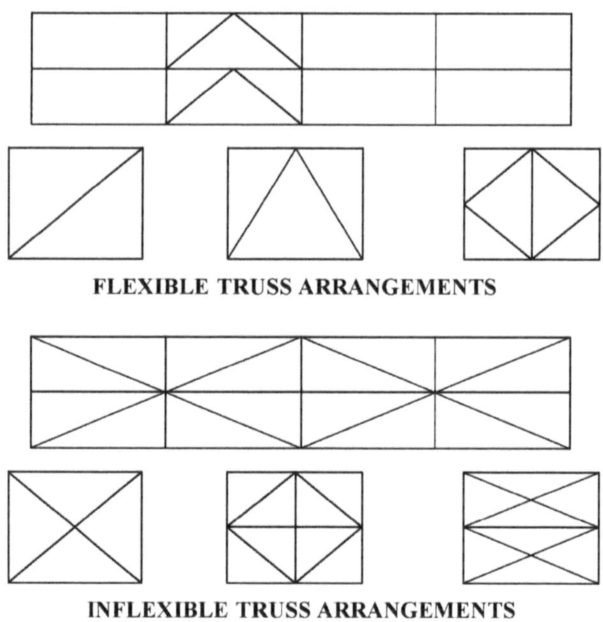

Figure 2-21. Typical flexible and inflexible truss arrangements.

should be located either at the centerline of the stiffeners or in the plane of the plate, depending on the assumed load path and the resultant eccentricities. Work points should also be placed at abrupt changes in the contour of the duct wall, such as corners or changes in slope. If loads are predominantly in one direction, the arrangement should put as many of the internal members in tension as possible, not only for structural economy but also to minimize the flow disruption, because tension members tend to be smaller than compression members. And, as with any truss, extremely small angles (less than 30°) between members should be avoided, because the connections become less compact and more difficult to design and build.

Because trusses and struts may affect the flow and pressure drop, their benefits should be weighed carefully before the project team makes final decisions on their use. The mechanical process engineer should be notified of any planned internal bracing so that the effect on the flow performance of the system can be assessed. Also, the structural engineer should decide whether the service environment inside the ducts is so poor as to limit the use of internal components such as trusses and struts. For example, it may be prudent to limit the use of internal trusses and struts in a flue-gas desulfurization system outlet duct, because the internal members would have to be fabricated from expensive corrosion-resistant alloys or encapsulated in some sort of protective material to survive in the harsh chemical environment.

Alternatives to internal trusses and struts include designing stiffener frames as moment-resistant frames, designing openings in duct walls as Vierendeel trusses, and using larger stiffeners that can span the full width of the duct. These alternatives will usually raise the ductwork's initial cost, because the greater weight of the stiffeners and more field welding at the frame moment connections will have a larger cost impact than the elimination of the trusses. These alternatives will also reduce the overall stiffness of the duct, which may be an important factor in certain applications.

2.6.2 Vibration Considerations

Another factor to be considered in the design of internal trusses, struts, and flow distribution devices is their susceptibility to flow-induced vibrations. This topic is covered in more detail in Chapter 8. To reduce the potential for vibrations, internal trusses or struts should be avoided immediately upstream or downstream of large fans. The exact requirements should be requested from the fan manufacturer. When the fan manufacturer does not furnish this information, Section 8 of Air Movement and Control Association's Document 201, *Fans and Systems* (2007) provides excellent guidance. This is discussed in more detail in Section 2.2. Where internal trusses or struts cannot be eliminated near large fans, their natural frequencies should be checked to avoid resonance with the fan, and the fan manufacturer should be consulted for any other possible adverse effects.

2.6.3 Erosion Considerations

The mechanical process engineer should identify regions in the duct arrangement that warrant erosion protection based on the abrasiveness of the flow medium and the gas velocities. Potential areas of concern involve internal trusses, struts, instruments, flow distribution devices, and duct walls. Physical or computational flow modeling can also be used to predict areas of high velocity or concentrated particulate flow. Sacrificial erosion-protection material may be included in the duct fabrication by increasing the thickness of the components or adding additional plate work. The material, thickness, and location of erosion protection to be included in the duct fabrication should be determined by mutual agreement between the supplier and the end user depending on the degree of protection desired.

Detailed design considerations for erosion protection include the material, geometry, and attachment method. In general, harder materials will exhibit better wear resistance, as long as their material properties are not degraded by the service temperature. Placement of erosion-protection plate normal to the gas flow provides optimal wear resistance. Possible differential thermal expansion of the erosion-protection material should be allowed for when determining its method of attachment to the duct components.

The structural members and walls of duct sections should not be fabricated from hardened abrasion-resistant materials, given their nonductile properties and the potential for fatigue cracking failures.

2.7 EFFECTS OF THE ARRANGEMENT ON LOADS

The ductwork arrangement will affect the loads that the duct and the support structure experience. These effects are discussed here. For more detailed information on the definition, determination, and application of loads, refer to Chapter 4. Chapter 5 outlines typical loading combinations.

2.7.1 Static Pressure

Static internal pressures are present in all ducts because of the fans that force the air or flue-gas through the system. Depending on the overall arrangement of the system and the location of the particular duct section within the system, these pressures may be positive or negative and will vary from section to section. Pressures will also vary under different operating modes.

Whenever a section of duct has an expansion joint located so that there is a duct wall that does not have another duct wall directly across from it, an unbalanced pressure occurs in that duct section. When this occurs, the resulting unbalanced pressure force must be transferred through the duct to its support structure. If no mechanism is available to resist the unbalanced pressure force, the duct section may move off its supports, which could cause considerable damage to the duct, the support structure, and nearby structures and equipment. Unbalanced pressure forces will only occur opposite openings or expansion joints directly upstream or downstream of bends or transitions. The structural engineer may exercise some level of control over these forces through judiciously orienting openings, improving the geometry of transition sections, keeping the duct runs as straight as possible, and ensuring the proper placement of dampers and expansion joints (Figure 2-5).

2.7.2 Dynamic Pressure

As air or flue-gas flows through the ductwork, it has momentum. The duct must exert a force to redirect that momentum at each bend in the duct. The magnitude of these forces depends on the velocity and density of the air or flue-gas and the angle of the bend. These forces are not usually significant compared to the other forces present, but it is the structural engineer's responsibility to determine whether this is true of the particular duct section being designed. If the dynamic pressure loads are significant, they are to be included in the design of the duct and its support structure.

Dynamic pressure forces are also developed as the air or flue-gas moves past internal flow distribution devices, such as turning vanes, perforated plates, and silencers. For more information on dynamic pressure loads, refer to Chapters 4 and 9.

2.7.3 Vibrations

Vibrations in ductwork may result from interaction with the equipment or may be induced by flow-related phenomena. Usually only steady vibrations are a concern.

The large fans in the duct system are the most common sources of equipment-related vibrations. The direct transmission of vibrations to the adjacent ducts from fan housings or other equipment can usually be avoided by isolating the equipment from the duct with expansion joints.

Flow-related vibrations can be caused by vortex shedding, gas pulsations, or resonance. Vortex shedding affects internal duct components, such as internal truss members and struts. Flow pulsations may occur close to fans or at the boiler outlet, where the duct may experience boiler-induced pulses. Flow pulsations will be synchronous with the fan blade tip frequencies and boiler combustion pulses. Isolating the duct with expansion joints will not alleviate flow-related vibrations.

Another potential source of vibration is ash collection hoppers. The flow of collected material through the hoppers can cause intermittent vibrations. Devices installed to help the material flow, such as rappers or air cannons, can also cause vibrations. These intermittent hopper vibrations are usually not of major concern.

To avoid vibration-related problems, flow distribution devices, internal trusses, and struts should be avoided immediately upstream and downstream of major fans. This is discussed further in Section 2.6. If installing these items within the duct cannot be avoided, the natural frequencies of the elements should be checked to avoid resonance. As discussed in Chapters 7 and 8, the spacing and sizing of stiffeners near major fans should also be carefully calculated to reduce or eliminate vibration concerns. For more information on designing flow distribution devices for vibrations, refer to Chapter 9.

2.7.4 Ash Deposition

Ash particles will drop out of the flow stream when the flue-gas velocity drops below a certain value. The exact value depends on the particle size, shape, and density. Experience and flow model tests have shown that for most fossil fuel ashes, drop-out should be within acceptable limits if the flue-gas velocity is kept above 40 ft/s (12 m/s). But even when the velocity is kept well above this value, it is common to find deposits of ash in the corners of ducts, especially at duct turns; at transitions from smaller ducts to larger ducts; and at obstacles to the flow, such as gusset plates for internal trusses. Ash deposition is also a concern in dead legs. The flue-gas is likely to become relatively stagnant in the dead leg, making particulate drop-out much more severe. Also, fly ash is often carried over from the flue-gas ducts to the air ducts through the air preheaters. When this is expected, the air ducts should be designed for the expected ash accumulation.

Ash deposition affects the structural design of ducts in two ways. First, the ducts are to be designed for the dead load of the ash. Second, if the ash accumulation is excessive, it will act as an insulator, which may increase the thermal gradients in the ducts. For more information on designing for ash deposition and thermal gradients, refer to Section 2.7.6 and Chapter 4.

Accumulated ash in expansion joints can harden and reduce the joints' effectiveness. Packed ash can also contribute to premature degradation of the joint material. The expansion joint manufacturer should be consulted regarding any design features that may be available to alleviate these concerns.

2.7.5 Metal Expansion Joints

If metal expansion joints are used in the duct system, they will affect the distribution of loads within and between duct sections and the equipment. This is because metal expansion joints have appreciable stiffnesses in all directions and can therefore transmit sizeable forces across the joints to the ducts on either side of the joints. A metal expansion joint's stiffness depends on the joint size, material, and design configuration and should be obtained directly from the joint manufacturer. Depending on the ductwork layout, shear stiffness of metal expansion joints may have a significant impact on the load distribution to the supports. The structural analysis may need to consider the effects of metal expansion joints especially when adjacent duct segments are supported at different elevations or where the stiffness of the metal expansion joint relative to the ductwork and duct supports may induce load transfer between adjacent duct segments.

2.7.6 Thermal Gradients

Thermal gradients across a section of ductwork or across an individual duct structural element can develop in any system; however, they are usually structurally significant only in hot ducts. Thermal gradients are typically most severe during unit start-up and shutdown conditions. If the temperature differentials are large, or if the ductwork is indeterminate, the gradients may introduce large stresses in the ducts and may cause a redistribution of all loads among the supports. In severe cases, the gradients can warp a duct section enough to cause some of its supports to lift up off the support structure. The uplifted supports are thus rendered nonfunctional, and the remaining supports and the support structure can become severely distressed as the loads redistribute.

The overall arrangement of the ductwork system can have a great influence on whether significant thermal gradients are likely to develop. Flow model testing can help identify problem areas and indicate where thermal gradients are likely to develop. The preferred approach is to avoid thermal gradients rather than to try to design for them.

Addressing the following items is important for minimizing thermal gradients, especially in hot ducts.

- As discussed in Section 2.2.3, dead legs should be minimized to the greatest extent practical.
- As discussed in Section 2.7.4, flue-gas velocity should be maintained above 40 ft/s (12 m/s) to minimize ash deposition, because excessive ash will act as an insulator.
- False floors, walls, or ceilings should be implemented with caution, because dead zones behind the falsework are not directly exposed to the hot flue-gas. The exposed walls will heat up or cool off faster than the hidden walls during start-up or shutdown. The dead air space may act as an insulator, creating a gradient even during steady-state operating conditions.
- Members that are directly in the path of the flue-gas stream, such as internal truss members, may heat up or cool off faster than the duct walls during unit start-up or shutdown. Truss arrangements that can accommodate the resulting thermal stresses should be used wherever possible, as discussed in Section 2.6.
- Poor flow distribution or stratification can create problems by exposing some portions of the ducts to flow streams of differing characteristics. Sections of ductwork where the mixing of two distinct flow streams takes place are usually vulnerable to stratification. Common examples include locations where ducts from two or more boilers merge into a common duct before entering an air quality control device or the chimney, tempering air is added to a duct, and bypass flue-gas reenters the ductwork downstream of a scrubber. Flow model testing can often help identify arrangements that are susceptible to stratification. Flow distribution devices, such as turning vanes, mixing plates, baffles, and perforated plates, may be added to alleviate the stratification.

Thermal gradients can also develop locally in duct plate, external stiffeners, expansion joint frames, duct support legs, and similar components, depending on the insulation and lagging details used. The typical design basis is that ductwork is not operated with insulation removed, to minimize the potential damage caused by thermal gradients, to ensure that local cold spots do not develop that may lead to corrosion of the duct plate, and to protect personnel. Refer to Chapter 11 for more on thermal gradients caused by poor insulation details.

Nagging structural problems or damage to existing ducts may arise from temperature differentials within the duct section. This effect can often be quantified by using thermocouples to collect temperature data from within the duct and then performing a structural analysis using these temperatures. This data collection is discussed in Chapter 12.

References

Air Movement and Control Association. 2007. *Fans and systems: Publication 201-02 (R2007)*. Arlington Heights, IL: Air Movement and Control Association.

ASCE Retrofit Task Committee. 1991. *Retrofitting fossil plant facilities: Structural perspectives*. New York: ASCE.

Fluid Sealing Association. 2010. *Ducting systems non-metallic expansion joint technical handbook*, 4th ed. Philadelphia: Fluid Sealing Association.

CHAPTER 3

Structural Material: Selection, Application, and Properties

3.1 INTRODUCTION

The structural design of air and gas ductwork requires that the appropriate material be selected for the anticipated operating and exposure conditions. The entire content of this book on the structural design of ducts assumes that some type of steel will be used for the ducts and all their structural elements. The selection of the proper steel is the responsibility of the structural engineer designing the ductwork. The steel design strengths are expected to be proportioned in accordance with the design procedures detailed in Chapters 5, 6, 7, 8, and 9.

The remainder of this chapter makes reference to a number of ASTM International material specifications and the Unified Numbering System (UNS) and associated alloy designations for steels that are commonly used in ductwork construction. Such references are not intended to preclude the use of other steels but rather to represent the materials currently most commonly used for ducts.

3.2 AVAILABILITY OF MATERIALS

The steels most commonly used for the fabrication of ductwork are typically widely available. The structural engineer should, however, verify the availability of the steel to be used in the design and adjust the design approach accordingly. Limitations exist on the available sizes of certain steels, such as ASTM A36, which is not generally available in thicknesses less than 3/16 in. (5 mm). Rolled shapes are also not available for some thicknesses and grades of certain steels, necessitating the use of built-up sections. Special consideration should also be given to the lead time for purchasing some rolled shapes. Hollow structural sections may also be more available in certain thicknesses than others.

Another concern for the structural engineer is the availability and lead time of certain alloys that are used in highly corrosive environments (e.g., wet scrubber ducts), particularly when the material is used in small quantities. These alloys are discussed in Section 3.4.3.

It should be noted that ASTM A242 is generally no longer specified for ductwork. It has been used in the past, and it continues to be discussed herein to provide the structural engineer with information for use in retrofit applications.

3.3 MATERIAL PROPERTIES

The structural design of ductwork must account for changes in the physical properties of the steel during the anticipated operating and excursion conditions. For ductwork in power plant and industrial boiler applications, the condition that has the greatest impact on the material properties is usually the temperature. Ductwork will typically be exposed to a wide, but predictable, range of temperature. Over its life, a duct will be exposed to approximately the same operating temperature range most of the time. The upper end of this temperature range is typically used as the design temperature for the duct, and the steel's design strength should be based on this temperature for all normal operating loading combinations, as explained in detail in Chapter 5.

The material properties of steel vary significantly with temperature. The structural engineer must determine the appropriate design values based on these reduced properties. The engineer should be cautious when using temperature-reduced material properties from external sources, because most test data and published values refer to short-term temperature events (such as fire), where large inelastic deformations are considered acceptable given the extreme rarity of such events. Typically the elastic modulus and yield strength are determined based on a 2% strain for these rare and short-term temperature events, similar to the values published in American Institute of Steel Construction (AISC) Design Guide 19 (2003) and Appendix 4 of AISC 360 (2017), instead of the standard 0.2% strain. This makes a significant difference, because at elevated temperatures the stress–strain curve loses its well-defined yield point and the curve becomes nonlinear at earlier stages of loading.

The properties of primary interest with regard to structural design are the coefficient of thermal expansion, the modulus of elasticity, yield strength, and ultimate strength. Depending on the type of steel, consideration should also be given to creep, temper embrittlement, and graphitization when exposure to high temperatures is anticipated. Everything presented in this book assumes that the structural engineer understands basic steel behavior and the proper application of the terms and values presented herein.

The values presented in the following subsections typify values that are often used by structural engineers in designing ductwork exposed to temperatures above ambient. The values presented in this section have been obtained from information published by ASTM International, the US Steel Corporation, and the American Society of Mechanical Engineers (ASME). These values should be assumed to be guidelines. The exact values are a function of the actual chemistry of the steel used. Those used for design should be confirmed by the structural

Table 3-1. Commonly Used Material Properties Designations.

Alloy designation	*UNS designation*
625	N06625
686	N06686
AL-6XN	N08376
Alloy 59	N06059
C-22	N06022
C-276	N10276
C-2000	N06200

engineer from one of the following sources: proprietary data; material test results; values published by AISC, ASME, or ASTM; or the steel supplier.

Alloy material properties listed in the following subsections reference their commonly used alloy designations. Table 3-1 provides the corresponding UNS numbers.

3.3.1 Coefficient of Thermal Expansion

The coefficient of thermal expansion of steel, ϵ, increases slightly with temperature. Figures 3-1 and 3-2 show the effect of temperature on ϵ for some of the carbon, stainless, and alloy steels commonly used in ductwork applications. The value used in the figures is the mean coefficient of thermal expansion (in./in./°F) as the steel temperature increases from 70 °F (20 °C) to the indicated temperature.

3.3.2 Modulus of Elasticity

The modulus of elasticity, E, is the ratio of stress to strain in the steel's elastic range. It decreases as the steel's temperature increases. Figure 3-3 shows the effect for some of the carbon, stainless, and alloy steels commonly used in ductwork applications. There are temperature limitations commonly imposed for the practical structural application of some of the steels shown in this graph. These limitations are discussed in Section 3.4.2.

3.3.3 Yield Strength

The yield strength of steel, F_y, also decreases with increasing steel temperature. A steel's yield strength may be quantified by the ratio of the elevated-temperature yield strength to the ambient-temperature yield strength, which is the value presented in the ASTM specifications. The effect of temperature on the yield strength ratio for some of the carbon, stainless, and alloy steels commonly used in ductwork applications are shown in Figures 3-4, 3-5, and 3-6, respectively. The design yield strength is arrived at by multiplying the ASTM published ambient value by the yield strength ratio.

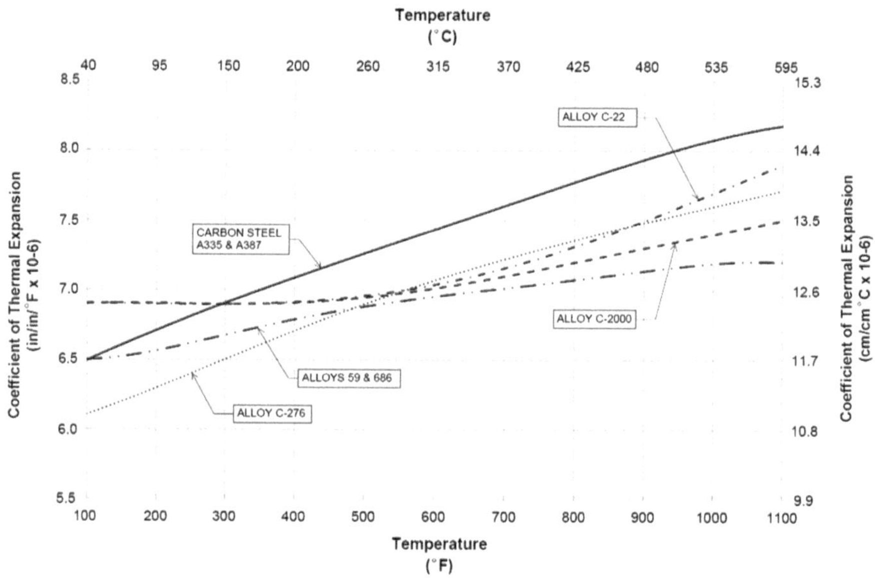

Figure 3-1. Coefficient of thermal expansion versus temperature for carbon steel and alloys.

The variation of yield strength with temperature is somewhat different for various grades and types of steel because of differences in their chemistry and microstructure. Temperature limitations are commonly imposed for the practical structural application of some of the steels shown on this graph. These limitations are discussed in Section 3.4.2.

3.3.4 Tensile Strength

The tensile strength of steel, F_u, also decreases with increasing steel temperature. It depends on the type of steel used but it may be quantified by the ratio of the elevated-temperature tensile strength to the ambient-temperature tensile strength. The effect of temperature on this tensile strength ratio for some of the carbon, stainless, and alloy steels commonly used in ductwork applications is shown in Figures 3-7, 3-8, and 3-9, respectively. The design tensile strength is arrived at by multiplying the ASTM published ambient value by the tensile strength ratio.

The variation of tensile strength with temperature is also somewhat different for various grades and types of steel because of differences in their chemistry and microstructure. Also, steels having relatively high percentages of carbon will exhibit strain aging in the range of 300 °F to 700 °F (150 °C to 370 °C), producing somewhat higher tensile strengths than those shown in Figure 3-7. This increase in tensile strength is generally not taken into account in practice. Temperature limitations are commonly imposed for the practical structural application of some of the steels shown in this graph. These limitations are discussed in Section 3.4.2.

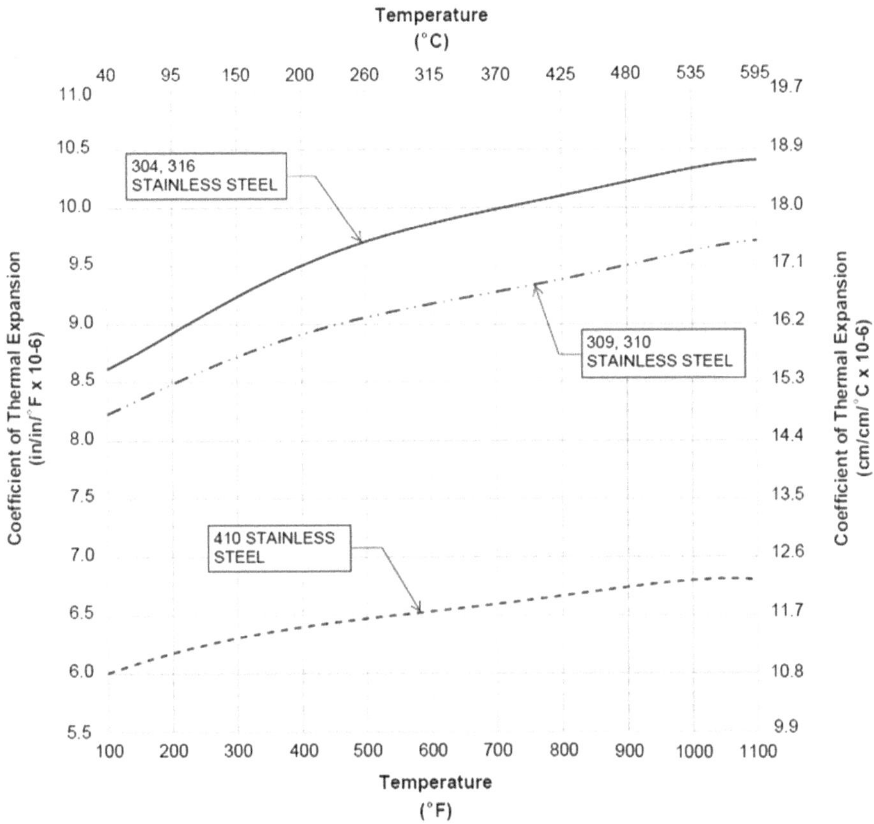

Figure 3-2. Coefficient of thermal expansion versus temperature for stainless steel.

3.3.5 Creep and Creep-Rupture

At high temperatures, steels exhibit creep, and their strength becomes strain-rate dependent. With a constant tensile stress, the steel strain will increase (creep) continuously as time progresses. Considerable strain and eventual failure may occur at stresses well below the yield strength as measured in the typical ASTM short-duration test. This is a creep-rupture failure, and the stress that causes a creep-rupture failure in a given time at a given temperature is the creep-rupture strength of the material, sometimes also referred to as the stress-for-rupture. The duration required to cause rupture is known as the time to failure. Depending on the stress level, the exact steel chemistry, and the exposure temperature, it can be a matter of hours or years. For relatively low stresses in normal design ranges, it is usually thousands of hours. Creep is distinguished from fatigue and fracture in that creep takes place within the grain boundaries, and fatigue is an intergranular occurrence.

For long-term high-temperature service, especially when the temperature is above 750 °F (400 °C), the creep-rupture strength of the steel becomes a primary

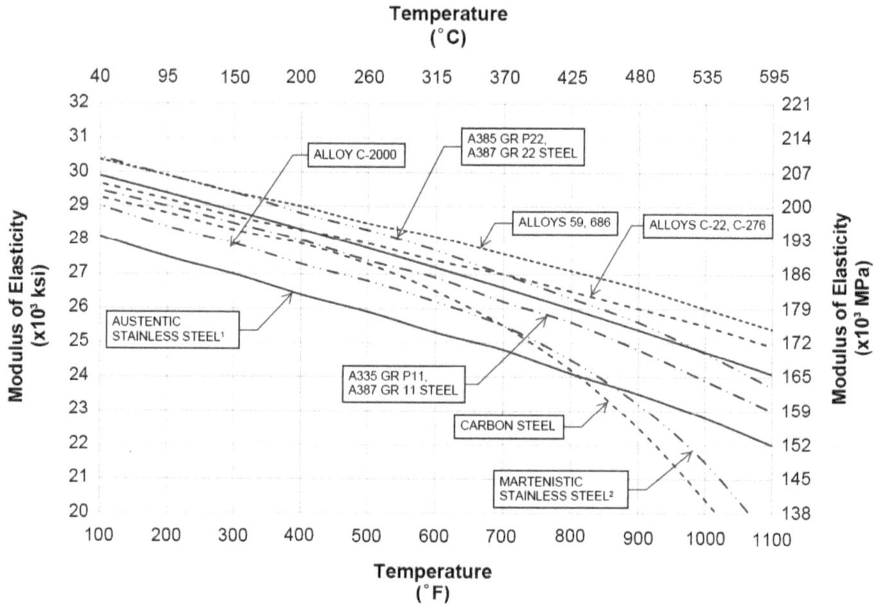

[1] Austenitic stainless steel, includes 304, 309, 310, and 316.

[2]Martenistic stainless steel, includes 410.

Figure 3-3. Modulus of elasticity versus temperature.

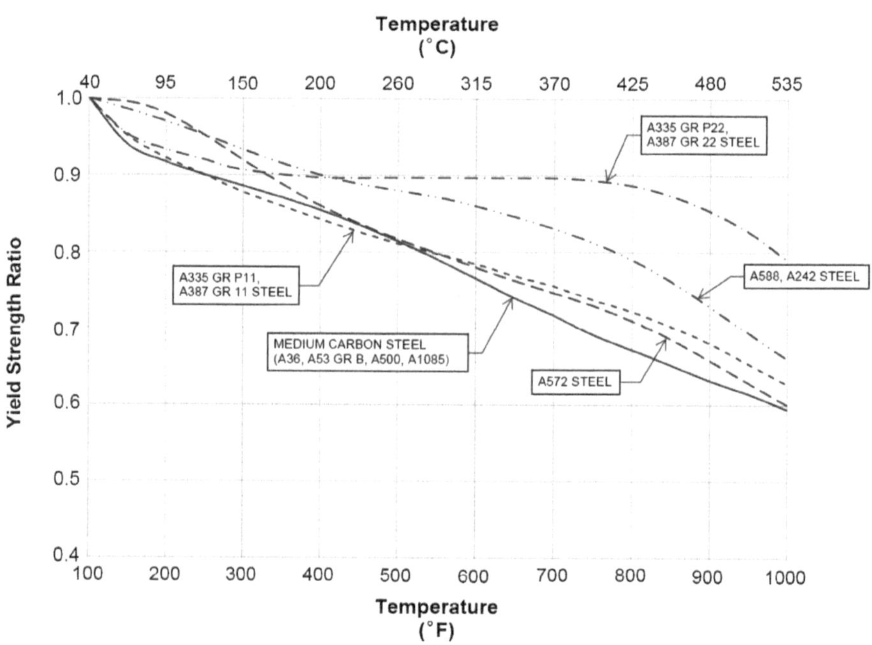

Figure 3-4. Yield strength ratio versus temperature for carbon steel.

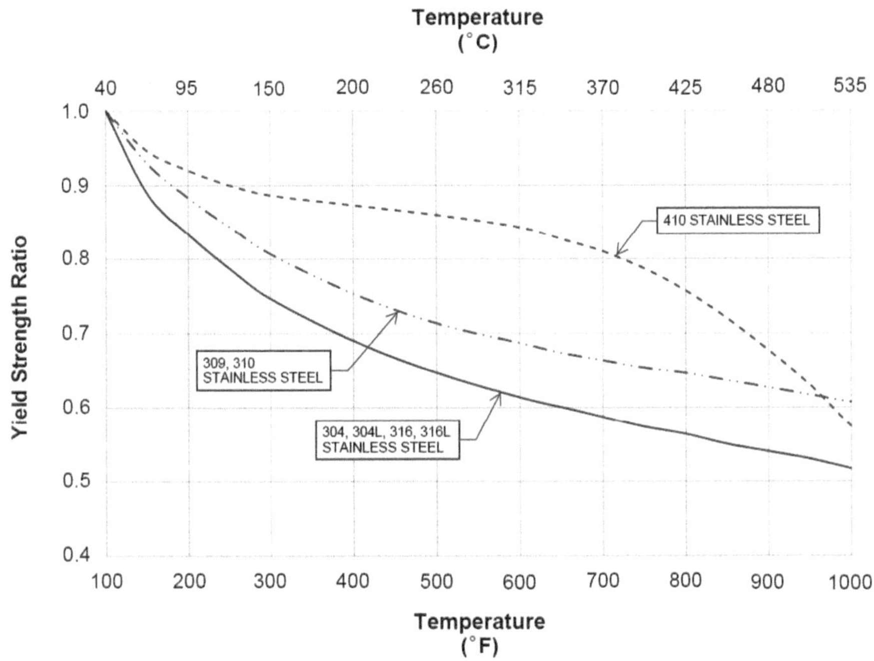

Figure 3-5. Yield strength ratio versus temperature for stainless steel.

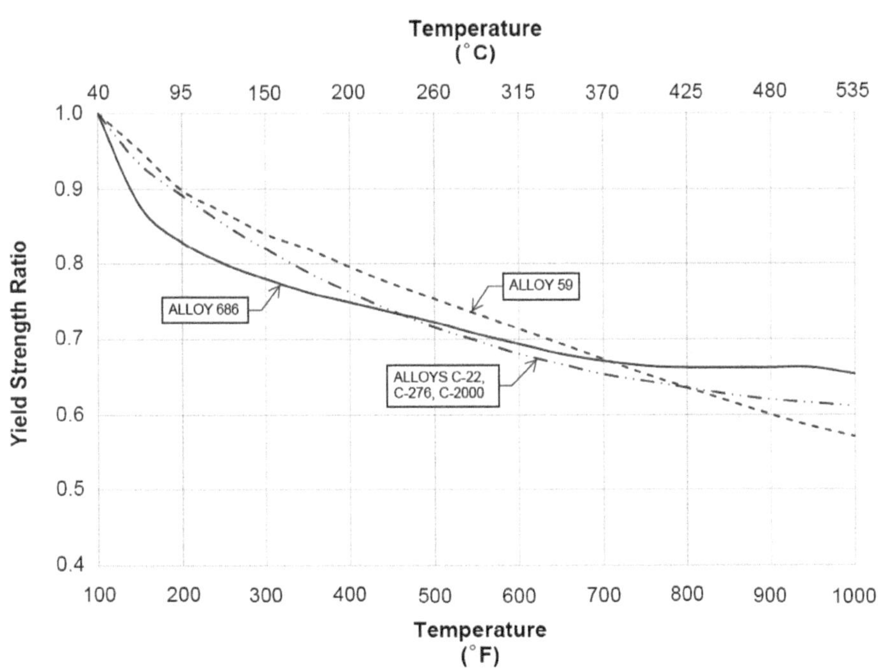

Figure 3-6. Yield strength ratio versus temperature for alloy steel.

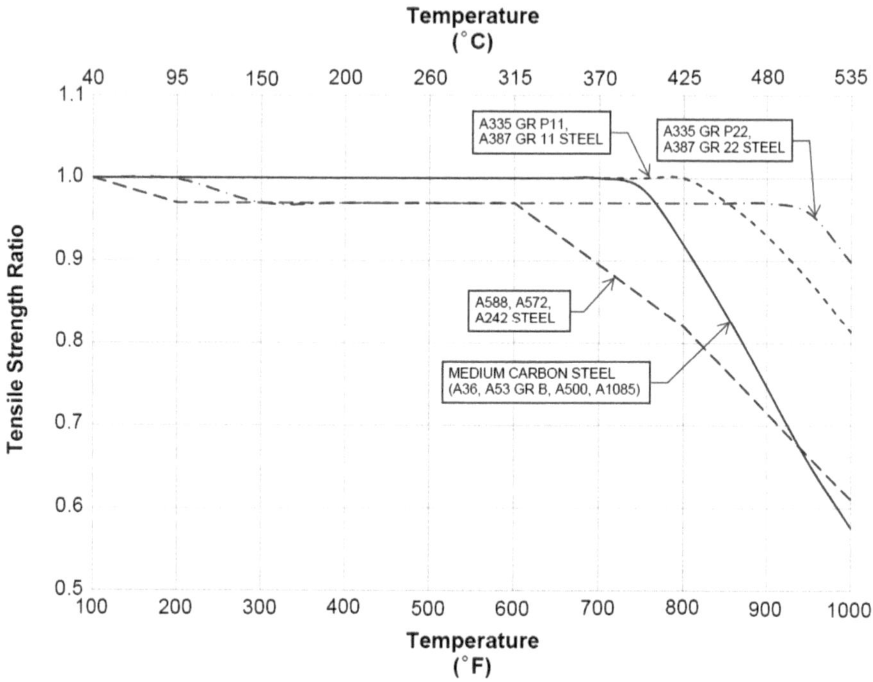

Figure 3-7. Tensile strength ratio versus temperature for carbon steel.

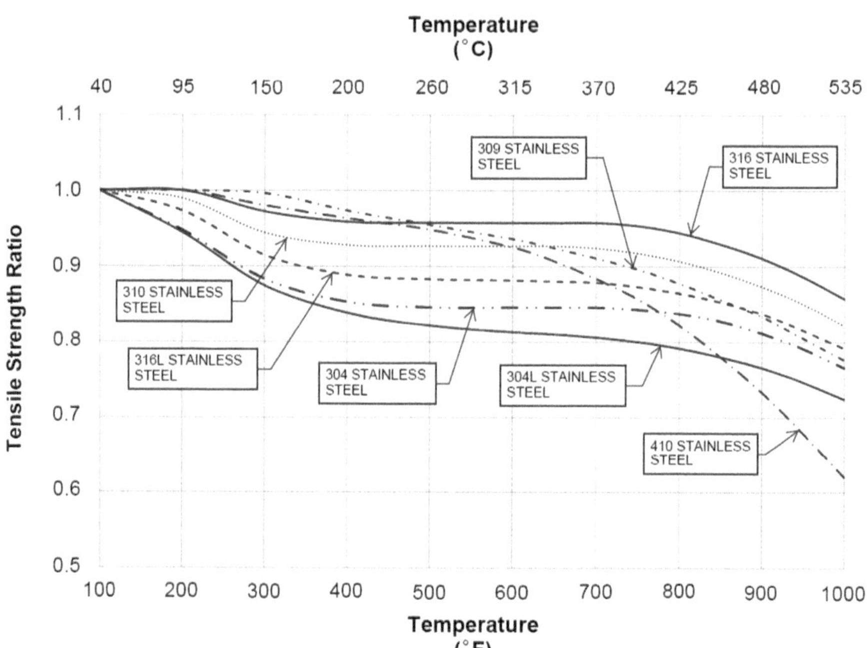

Figure 3-8. Tensile strength ratio versus temperature for stainless steel.

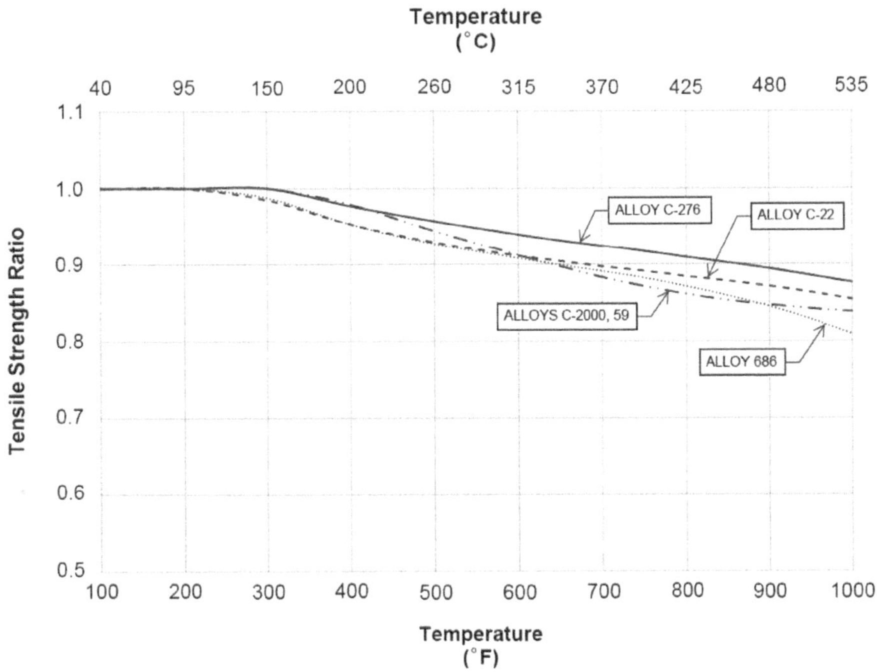

Figure 3-9. Tensile strength ratio versus temperature for alloy steel.

factor in determining the design strength. A limiting creep rate should also be considered in establishing the design strength. Creep and creep-rupture strength vary inversely with temperature and with the duration of exposure to that temperature.

Creep and creep-rupture also strongly depend on the exact chemistry of the steel. Some steels, such as ASTM A36, A572, and A992, are very susceptible to creep and creep-rupture; others, such as austenitic (Cr-Ni) stainless steels, are almost creep resistant with operating temperatures at or below 1,000 °F (540 °C). Although the ASTM specification to which a steel conforms helps qualify its susceptibility to creep and rupture, the exact chemical composition of the steel is necessary to quantify its creep and creep-rupture properties.

The design creep life should be selected by the structural engineer based on design criteria consistent with the expected service life and service conditions. Creep-rupture allowable stress values published in the ASME *Boiler and Pressure Vessel Code* are based on a 100,000 h creep life. A shorter or longer creep design life may be appropriate depending on the operating mode and the expected service life of the ductwork. Creep rupture stress values for a shorter or longer duration creep life may be determined by using the ASME published values or the values in Table 3-2 and adjusting these using the Larson-Miller parameter. For more information on the factors used to limit allowable creep stresses, see the ASME *Boiler and Pressure Vessel Code* (2015a) Section II– *Materials, Part D, Mandatory Appendix 1.*

Table 3-2. Typical Creep Stresses (ksi) for Steels Common in Ductwork Applications.

		Temperature									
Material	*°F (°C)*	*750 (400)*	*800 (415)*	*850 (455)*	*900 (480)*	*950 (510)*	*1000 (535)*	*1050 (565)*	*1100 (595)*	*1200 (649)*	*Notes*
A36	Creep Rate Stress[1]	–	14.0	9.1	5.9	3.9	2.6	1.6	1.1	–	See Note 4
A53 Gr. B	ksi (MPa)		(96.5)	(62.7)	(40.7)	(26.9)	(17.9)	(11)	(7.6)		
	Creep Rupture Stress[2]	–	16.0	11.2	8.2	5.8	4.2	2.9	2.2	–	
	ksi (MPa)		(110.3)	(77.2)	(56.5)	(40)	(29)	(20)	(15.2)		
A242 Type 1	Creep Rate Stress[1]	–	39.1	–	17.7	–	9.0	–	–	1.0	See Note 5
A618 Gr 1	ksi (MPa)		(269.6)		(122)		(62.1)			(6.9)	
	Creep Rupture Stress[2]	–	41.6	–	19.1	–	10.0	–	–	1.0	
	ksi (MPa)		(286.8)		(131.7)		(68.9)			(6.9)	
A588 Gr A	Creep Rate Stress[1]	–	29.2	–	16.3	–	8.6	–	–	1.0	See Note 6
A588 Gr B	ksi (MPa)		(201.3)		(112.4)		(59.3)			(6.9)	
	Creep Rupture Stress[2]	–	28.9	–	17.3	–	8.3	–	–	1.0	
	ksi (MPa)		(199.3)		(119.3)		(57.2)			(6.9)	
A387 Gr 11	Creep Rate Stress[1]	–[3]	–[3]	23.7	16.7	11.3	7.5	4.8	2.8	1.2	See Note 7
Class 2 Plate	ksi (MPa)			(163.4)	(115.1)	(77.9)	(51.7)	(33.1)	(19.3)	(8.3)	
	Creep Rupture Stress[2]	–[3]	–[3]	33.6	23.8	16.3	11	7.5	–	–	
	ksi (MPa)			(231.7)	(164.1)	(112.4)	(75.8)	(51.7)			
A335 P11	Creep Rate Stress[1]	–[3]	–[3]	–[3]	–[3]	11.3	7.5	4.8	2.8	1.2	See Note 7
	ksi (MPa)					(77.9)	(51.7)	(33.1)	(19.3)	(8.3)	
	Creep Rupture Stress[2]	–[3]	–[3]	–[3]	–[3]	16.3	11.0	7.5	–	–	
	ksi (MPa)					(112.4)	(75.8)	(51.7)			
A387 Gr 22	Creep Rate Stress[1]	–[3]	–[3]	–[3]	–	–	7.3	5.8	4.0	1.2	See Note 7
Class 2 Plate	ksi (MPa)						(50.3)	(40)	(27.6)	(8.3)	
	Creep Rupture Stress[2]	–[3]	–[3]	–[3]	17.0	14.0	10.0	7.3	5.2	1.7	
	ksi (MPa)				(117.2)	(96.5)	(68.9)	(50.3)	(35.9)	(11.7)	

A335 P22	Creep Rate Stress[1]	–[3]	–[3]	–[3]	–	–	7.3	5.8	4.0	1.0	See Note 7
	ksi (MPa)						(50.3)	(40)	(27.6)	(6.9)	
	Creep Rupture Stress[2]	–[3]	–[3]	–[3]	17.0	14.0	10.0	7.3	5.2	1.7	
	ksi (MPa)				(117.2)	(96.5)	(68.9)	(50.3)	(35.9)	(11.7)	
304L	Creep Rate Stress[1]	–	–	–	–	–	7.8[9]	6.3[9]	5.1	3.3	See Note 8
	ksi (MPa)						(53.8)	(43.4)	(35.2)	(22.4)	
	Creep Rupture Stress[2]	–	–	–	–	–	19.5[9]	15.0	11.6	6.9	
	ksi (MPa)						(134.8)	(103.4)	(80)	(47.6)	
304/304H	Creep Rate Stress[1]	–	–	–	–	–	17.9	14.0	11.1	7.2	See Note 8
	ksi (MPa)						(123.4)	(96.5)	(76.5)	(49.6)	
	Creep Rupture Stress[2]	–	–	–	–	–	29.0	22.6	17.8	11.0	
	ksi (MPa)						(199.9)	(155.8)	(122.7)	(76.1)	
316L	Creep Rate Stress[1]	–	–	–	–	–	22.5	16.2	12.0	6.4	See Note 8
	ksi (MPa)						(155.1)	(111.7)	(82.7)	(44.1)	
	Creep Rupture Stress[2]	–	–	–	–	–	34.5[9]	25.0	18.5	10.1	
	ksi (MPa)						(237.9)	(172.4)	(127.6)	(69.6)	
316/316H	Creep Rate Stress[1]	–	–	–	–	–	20.1	15.8	12.4	7.9	See Note 8
	ksi (MPa)						(138.6)	(108.9)	(85.5)	(54.5)	
	Creep Rupture Stress[2]	–	–	–	–	–	37[9]	28.5	21.2	11.6	
	ksi (MPa)						(2613.1)	(196.5)	(146.2)	(80)	

Notes:[1]·Creep rate stress values are the average stress values needed to produce a creep rate of 1% in 100,000 hours of service with a factor of safety of 1.0.[2]·Creep rupture stress values are the average stress values needed to cause rupture after 100,000 hours of service with a factor of safety of 1.0.[3]·Where indicated, values are not time dependent at indicated temperatures.[4]·Values derived from ASME Boiler and Pressure Vessel Code, Section IID, with safety factors removed and from ASTM DS11S1. Note these steels recommended for noncritical applications above 750 °F (400 °C) and not recommended for use above 800 °F (415 °C).[5]·Source: United States Steel, *Steels for Elevated Temperature Service*. Recommended for noncritical applications above 900°F (480°C) and not recommended for use above 1,000 °F (535 °C).[6]·Source: *Structural Steel Designer's Handbook* by Brockenbrough and Merrit, 2nd Edition, and United States Steel, *Steels for Elevated Temerpature Service.*[7]·Values derived from ASME Boiler and Pressure Vessel Code, Section IID, with safety factors removed, ASTM DS 50 and ASTM DS6S2.[8]·Source: ASTM DS 5S2, *An Evaluation of the Yield, Tensile, Creep, and Rupture Strengths of Wrought 304, 316, 321, and 347 Stainless Steels at Elevated Temperatures*, 1969.[9]·Values extrapolated from ASTM DS 5S2 data series information.

Because of the nature of creep, the loading demand should be computed from load combinations that will be sustained at elevated temperatures. The element load demand arising from such load combinations should be below the creep strength. See Chapter 5 for more information on loading combinations with sustained loads. The maximum allowable creep design strength can be obtained from the allowable stress values published in *ASME Boiler and Pressure Vessel Code* (2015a) or other suitable resources.

Selected creep stresses (1% creep rate in 100,000 h) and rupture stresses (rupture in 100,000 h) for various steels are presented in Table 3-2. The values in this table represent the actual failure stresses (in the case of rupture) or 1% creep rate in 100,000 h, without any safety factors, from various sources.

Few or no published creep-rupture data are available for many steels. In Table 3-2, the published data for ASTM A588 are for Grade A. ASTM A588 Grade B has been included because of the belief that given the similar chemical properties, its creep-rupture behavior is similar to Grade A. It is also the Task Committee's belief that the creep-rupture properties in Table 3-2 for ASTM A53 Grade B are applicable to ASTM A500 and ASTM A1085 materials based on their chemical similarities. Creep-rupture data could not be found for other common steel specifications, such as A572 and A992. For all steels, the values in Table 3-2 are intended only as a reference. It is the structural engineer's responsibility to select final project design values for the creep rate strength at the applicable creep rate and the rupture strength design value. Values used in design should be obtained from test data reflecting the precise chemical composition of the steel to be used in the ductwork fabrication, when possible. If steels other than those in Table 3-2 are used at highly creep-susceptible temperatures, the structural engineer is responsible for determining the steel's creep and creep-rupture strength.

3.3.6 Temper Embrittlement and Graphitization

Steel also experiences other metallurgical changes at high temperatures that affect its mechanical properties, producing brittle characteristics that should be considered in the structural design. Those of importance in the structural design of ductwork include temper embrittlement and graphitization.

Temper embrittlement, sometimes referred to as *885 embrittlement*, occurs when certain steels are isothermally aged in the temperature range of 700 °F to 1,100 °F (370 °C to 600 °C). It is attributed to the segregation of impurities to the grain boundaries, and it is retarded by the presence of molybdenum, which tends to inhibit such segregation. Steels with relatively high percentages of phosphorus and traces of certain uncontrolled impurity elements such as tin, arsenic, and antimony are highly susceptible to temper embrittlement. Nickel, when in the presence of these impurities or phosphorus, also promotes temper embrittlement. Other elements common in steel chemistry, such as manganese and chromium, cause more molybdenum to be needed to minimize embrittlement. Materials with high chromium content (greater than 12%), such as ferritic, duplex, and martensitic stainless steels, tend to be highly susceptible to embrittlement.

When the steel becomes temper embrittled, it loses its toughness and easily cracks when stressed or impact loaded in the presence of a flaw, such as at a weld. Temper embrittlement raises the steel's nil ductility temperature from approximately −60 °F to 0 °F (−50 °C to −20 °C) to as much as 250 °F (120 °C). This causes the steel to be extremely susceptible to brittle fracture at ambient temperatures or during unit start-up and shutdown. Temper embrittlement effects can practically only be mitigated by specifying a steel chemistry that is not susceptible to temper embrittlement when it is to be exposed to these high temperatures. Steels that have temper embrittlement can have their original toughness restored by a heat treatment process, but for ductwork this is typically not practical given the size and location limitations.

Graphitization, sometimes called carbon migration, occurs in steels containing a relatively high amount of carbon. In this process graphite forms at the grain boundaries when the steel is exposed to temperatures over 800 °F (425 °C) for a prolonged period. Graphitization causes the steel to prematurely fail at the weakened grain boundaries. Graphitization is reduced by lowering the carbon content of the steel and by adding chromium.

If graphitization is detected early, the steel can be normalized and tempered just below the lower critical temperature to return the steel to its original microstructure. Steel that has experienced more severe graphitization should have the affected areas removed and replaced.

3.4 MATERIAL SELECTION

The selection of the type of steel to be used for a duct section should consider the service conditions to which the material will be exposed. The following subsections identify and discuss materials commonly used in various ductwork exposure conditions. The ASTM specifications discussed in this section are generally those for which plate, structural shapes, and bar products are available. Similar ASTM specifications exist for the thinner sheet and strip products. For example, ASTM A570 is the sheet and strip counterpart to ASTM A36 plate, and ASTM A606 is the sheet and strip counterpart to ASTM A242 and A588 plate.

3.4.1 Low-Temperature and Medium-Temperature Ducts

This subsection addresses material for ducts having an operating temperature of less than 750 °F (400 °C). Carbon steel is typically used for duct plates, stiffeners, and other rolled shapes in this temperature range. ASTM A36, A992 (stiffeners), and ASTM A572 are the most common, although high-strength low-alloy steels such as ASTM A588 Grade B may be used for better resistance to atmospheric corrosion (although better corrosion resistance in an environment other than atmospheric has not been documented). ASTM A572, A588, and A242 are higher-strength steels, so they would provide more strength, typically less overall weight, and perhaps more economy.

Internal bracing for this temperature range will generally be carbon steel as well. Typically, ASTM A53 Grade B, A500, A1085, A501, or A618 Grade 1 pipe sections are used for the internal brace members. Connection material is usually made from the same steel as the duct plate. (ASTM A1085 is a new material designation, and specific test data are not available at the time of this publication. Given its material similarity to ASTM A53, this book treats the two in a similar manner, but the engineer is responsible for verifying any strength modification values used in design.)

The material properties of steel in this category should be adjusted from the published design values, as indicated in Figures 3-1 through 3-9, when the ductwork is exposed to operating temperatures exceeding 100 °F (40 °C).

3.4.2 High-Temperature Ducts

This subsection addresses material for ducts having an operating temperature greater than 750 °F (400 °C) but less than 1,100 °F (600 °C). In this temperature range, many steels reach a temperature limit where alternative steels become more attractive. The selection of a material for a given temperature requires judgment and is normally based on experience or the interpretation of available material test results. The following guidelines reflect the current industry practice and are intended to assist in the selection of a high-temperature material for ductwork plates and shapes.

ASTM A36, A572, and A992 steels have limited uses at temperatures above 750 °F (400 °C). Creep and graphitization become more predominant above this temperature. The use of ASTM A36 and A572 steel above 800 °F (430 °C) is definitely not recommended.

ASTM A242 Type 1 steel has limited uses between 900 °F and 1,000 °F (480 °C and 540 °C). When the steel's temperature exceeds 900 °F (480 °C), there is a significant drop-off in creep-rupture strength. Therefore, above 900 °F (480 °C), ASTM A242 steel is not recommended for structural applications unless the structural engineer takes steps to alleviate these concerns. At temperatures above 900 °F (480 °C), the structural engineer should investigate a better performing material having higher creep-rupture capacity and less susceptibility to temper embrittlement. This steel is susceptible to temper embrittlement when exposed to very high temperatures. More stringent control of the chemistry of this steel can dramatically improve its creep-rupture capacities but can also increase its susceptibility to temper embrittlement. The use of this steel above 1,000 °F (540 °C) is not recommended. It should be noted that this steel is rarely specified in current practice, and availability is questionable, but this material may be encountered in retrofit applications.

ASTM A588 Grade B has relatively high elevated-temperature strength, but low creep-rupture ductility above 800 °F (430 °C). Thus, it is normally not used in structural applications above this temperature. This steel may be used above 800 °F (430 °C) by carefully controlling and refining its relatively broad ASTM chemical composition range, specifically by limiting the vanadium inclusion to reduce embrittlement concerns. The use of this steel above 1,000 °F (540 °C) is definitely not recommended.

ASTM A335 and A387 can be used in various grades for exposure temperatures between 750 °F and 1,100 °F (400 °C and 600 °C).

Some of the ferritic and austenitic stainless steels have also been used and continue to be used in hot duct applications. Stainless steel has been used successfully for floating liners in combustion turbine exhaust ducts.

Internal bracing in high-temperature ductwork is generally made from ASTM A335 Grade P11 pipe section. The connection material will generally match the duct plate, but ASTM A387 may be used. The material properties for these steels, and any others, should be adjusted from the published design values, as indicated in Figures 3-1 through 3-7. Other considerations for high-temperature phenomena discussed in Section 3.3 should also be considered when using some of these steels.

All steels will exhibit oxidation and scaling when continuously exposed to temperatures above 950 °F (510 °C). Because of this effect, a corrosion allowance may be prudently included in the design of duct steel above this temperature.

3.4.3 Wet Ducts

This subsection addresses low-temperature ductwork that will be exposed to wet conditions such as are encountered in the presence of a wet FGD system (scrubber). Hot flue-gas typically enters the inlet of a scrubber at 250 °F to 450 °F (120 °C to 230 °C) and is cooled to 120 °F to 140 °F (50 °C to 60 °C) while passing through the scrubber. Some systems use a reheat process to raise the flue-gas temperature to around 200 °F (95 °C) to prevent condensation and thus corrosion. Protected or unprotected ASTM A36 or A572 carbon steel is commonly used in applications where a biomass fuel with no sulfur content, such as wood or peanut hulls, is used.

High-strength, low-alloy steels such as ASTM A588 and A242 are not recommended for use in wet ducts because the surface oxidizing properties of these steels cause the material to erode and corrode rather quickly.

High-alloy liners or special coatings are typically used when municipal or hazardous waste or coal with high chlorides is used for fuel. Austenitic stainless steels are commonly used in systems fueled with petroleum or coal, where chloride concentrations are low. These austenitic stainless steels include types 304L, 316L, 317L, and 317LMN per ASTM A240, and 904L per ASTM B673. "Super austenitic" stainless steels, such as AL6XN, may be used where typical austenitic stainless steels are too weak or where the operating environment has a high chloride or acid content.

Other high-nickel alloys, such as C-22, C-276, and C-2000, may be used in highly acidic operating environments, such as downstream from wet FGD systems. These materials have been developed to maintain high strength at elevated-temperature operating conditions while resisting corrosion by flue-gas acidity; however, their cost can be prohibitive.

In addition to temperature, material selection depends on the expected chemical composition of the scrubbed or unscrubbed flue-gas. Important corrosion factors are the flue-gas acid concentrations, the presence of chlorides or other corrosive chemicals, and the presence of solid corrosive deposits on the duct floor. ASTM G48 (2011), *Standard Specification for Test Method for Pitting and Crevice Corrosion Resistance of Stainless Steels and Related Alloys by the Use of Ferric Chloride Solutions*, presents procedures for determining the lowest temperature at which corrosion occurs in the material. The mechanical process engineer should be consulted when choosing the material for wet ducts in a relatively cool, corrosive environment.

In highly corrosive areas where chlorides or acids exist, high-alloy material is usually used. Nickel-based alloys with significant chromium and molybdenum added for resistance to pitting and crevice corrosion include ASTM B564 Alloy 625 and B575 Alloys C-22 and C-276. These materials are used as solid plate, as part of a clad plate, and in thin liner applications. The economics of solid alloy plate versus alloy-lined carbon steel should be considered when selecting the material for wet ducts. Before making any decision, the material suppliers should always be contacted regarding availability and costs.

Although the details are outside of the scope of this book, wet ducts can be designed and constructed from fiberglass-reinforced plastic, if the operating and excursion temperatures are low enough. It is highly resistant to corrosive environments. Designs are typically performed by the manufacturer in accordance with ASME RTP-1 (2013).

3.5 BOLTS

Typical ductwork construction uses welding to produce an airtight product capable of pressurization. Bolts will therefore have only limited applications. Where bolts are used for structural connections in ducts with operating temperatures under 650 °F (400 °C), they typically will be ASTM F3125 Grade A325 Type 1, with A563 Grade C nuts and F436 hardened steel washers. For structural connections in ducts with operating temperatures over 650 °F (400 °C), ASTM A193 Grade B7 or other appropriate stainless steel bolts are typically used, with ASTM A194 nuts. As with the duct steel, consideration must be given to the lower strength at high temperatures. The effect of temperature on the tensile strength ratio for some of the bolt materials commonly used in ductwork applications is shown in Figure 3-10. ASTM A307 bolts are generally specified for ductwork applications that are not subjected to high temperatures or significant loading. They are sometimes used as expansion joint frame connection bolts or erection fit-up bolts for welded field splices.

The use of galvanized bolts is not uncommon for ductwork external applications at low temperatures.

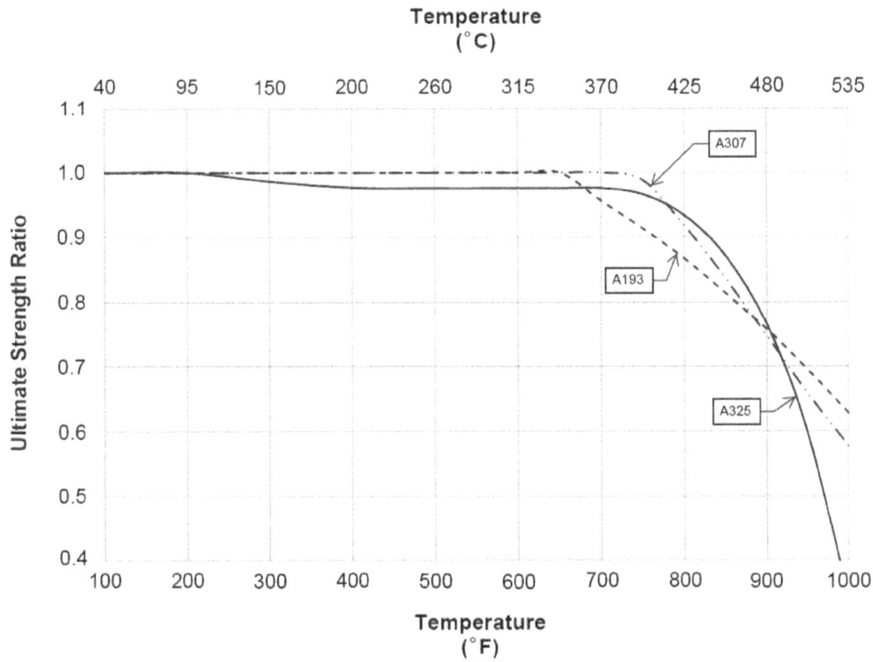

Figure 3-10. Tensile strength ratio versus temperature for bolts.

3.6 WELDING ELECTRODES

AWS D1.1 (2015), *Structural Welding Code, Steel,* is usually specified for ductwork structural welding. Sometimes the ASME *Boiler and Pressure Vessel Code* (2015b) Section IX, *Qualification Standard for Welding and Brazing Procedures, Welders, Brazers, and Welding and Brazing Operators,* is specified. For more on the use of these codes, and other welding information, see Section 10.3.

Welding electrodes with a minimum tensile strength of 70 ksi are commonly specified for carbon steel applications in ductwork construction. For high-temperature applications, above 750 °F (400 °C), 80 ksi electrodes are typically specified. Alloy steels, such as ASTM A335 and A387, generally use E8018-B2L electrodes, with welding conforming to AWS D10.8 (1996), *Recommended Practice for Welding of Chromium-Molybdenum Steel Piping and Tubing.* The appropriate electrodes must be selected when welds are made between dissimilar metals. The type of rod specified should be based on the higher-grade material being welded.

Due consideration must also be given to the reduction in weld metal strength when exposed to high temperatures. The temperature-based strength reductions for weld metal are usually assumed to be the same as for the base metal.

3.7 DUCTWORK PROTECTION

The exposed surfaces of a duct section are likely to be subjected to atmospheric conditions at various times during its service life. Many factors impact the selection of an appropriate surface protection system. The selection should include consideration of the susceptibility of the ductwork steel to corrosion. Other factors to consider are the flue-gas composition, operating temperature, and time of exposure to corrosive elements. The protection system chosen may consist of simply including a corrosion thickness allowance for the duct steel, or it may involve coating or placing a lining on the inside surface of the duct.

3.7.1 Corrosion Allowances

Ductwork composed of carbon steel may include a corrosion allowance. EN 13084-7 (European Committee for Standardization 2103, Table 4) provides additional guidance on corrosion allowance for surfaces in contact with flue-gas. Localized corrosion is typical at cold spots such as penetrations, expansion joints, and access doors, but these are typically small enough that the overall structural strength of the duct is unimpaired. In most situations, this localized corrosion does not warrant a corrosion allowance over the entire duct because stresses can generally redistribute around the areas exhibiting material loss. Owner input may need to be sought on the use of a corrosion allowance to reduce future maintenance costs.

It should also be noted that though it is not technically corrosion, highly abrasive ash conveyed in ducts at high velocities may erode internal structural members and flow-correction devices. Providing an erosion allowance can be one way of addressing this issue; another is to use sacrificial elements or abrasion-resistant materials, such as angles or plates welded to the upstream portion of pipe truss members.

A corrosion (or erosion) allowance is implemented by providing thicker material than required by design. Atmospheric-corrosion-resistant steels as ASTM A588 and A242 may not have better resistance than ASTM A36, A992, or A572 steel, because they rely on a tightly adhered layer of rust to seal the surface from oxygen. In a chemical environment, this layer of rust can be continually removed, causing high corrosion rates. High-temperature ductwork of alloy material may require no further consideration for corrosion. Insulation and lagging may usually be considered to afford sufficient protection against external corrosion. See Chapter 11 for more information on insulation and lagging.

3.7.2 Coatings

External coatings. Selection of a painting system for external ductwork surfaces requires careful evaluation of the anticipated environment and the long-term goal of the painting system. Painting of ductwork is of two kinds: protection for the life of the ductwork and protection during shipment, storage, and construction.

If the ductwork is expected to have low surface temperatures and will not be insulated and lagged, the exposed surfaces are typically painted with the same coating system used for the adjacent structural steel. If the ductwork will be insulated and lagged, painting might be used to protect the surfaces during shipment, storage, and construction. Then the appropriate primer would be an inexpensive shop primer, which could be classified as the manufacturer's standard. Sometimes a weldable primer is used. This primer will most likely burn off during operation, but the surfaces will remain protected by the insulation and lagging. In the rare case that ductwork is designed for elevated surface temperatures but is not insulated and lagged, a high-temperature coating is required. In any case, the compatibility of coating systems with the expected environment should always be investigated.

Consideration should be given to test ports, support legs, and other exposed items that penetrate the insulation and lagging. Penetrations can bring the temperature at their duct plate connections to corrosive flue-gas dew point levels. This is discussed further in Section 11.2. These items may experience elevated temperatures and should be treated with an appropriate heat-resistant coating system, or else they should be fabricated from corrosion-resistant materials. If the duct support legs are uncoated weathering steel, such as ASTM A242 or A588, the coated support steel will become stained or discolored from the "bleeding" of the weathering steel. This unsightly problem can be alleviated by coating the exposed weathering steel.

Internal coatings. Some FGD absorber outlet ducts are coated with a system designed to protect the duct steel from the harsh chemical environment. This system would be in lieu of a liner as discussed in Section 3.7.3. Further discussion of these types of internal coatings is beyond the scope of this book.

3.7.3 Liners

Ductwork liners are used to protect structural ductwork components from corrosion and erosion on the interior surfaces of the duct. Depending on the composition of the flue-gas, the inside surfaces of ductwork can be exposed to abrasive particulate and corrosive condensate, particularly as the temperature on the inside surface of the ductwork falls below the dew point of the flue-gas. The selection of a liner should consider flue-gas composition, operating temperature, and time of exposure to corrosive or erosive elements. The use of a liner results in a more expensive design and should therefore be limited to applications with harsh exposure conditions. Liner materials that are commonly used include brick, mortar, castable refractories, borosilicate glass blocks, rubber, and various steel and high-nickel alloys.

Metal liners may be stainless steels or high-nickel alloys. These liners are made from thin sheet metal placed like wallpaper on the interior surfaces of the ductwork. The liners are usually welded to the underlying steel, but they may be attached through a bonding process to form a clad plate. When liners are used, special care and detailing are required to ensure complete envelopment of the flue-gas by the corrosion-resistant and erosion-resistant liner material, particularly at welds, splices, corners, internal bracing, flow distribution device connections, and other breaks in continuity of the ductwork interior. A key factor in the successful installation of a metal liner is the welding. The welding electrode filler metal must correctly match the alloy base metal in corrosion resistance and mechanical properties. Special care should be taken during fabrication, shipment, and installation to avoid contamination of the liner material by carbon steel, especially at the welds between the carbon steel and the liner. Consideration should also be given to possible differences in coefficients of expansion and flexibility between the liner material and the underlying duct steel. More information on surface preparation, installation, and weld sequencing for thin metallic alloy "wallpaper" linings can be found in National Association of Corrosion Engineers (NACE 2012) SP 0292, *Installation of Thin Metallic Wallpaper Lining in Air Pollution Control and Other Process Equipment.* Liners are generally not considered to contribute to the duct's load carrying capability in the stress calculations except when absolutely necessary in retrofit applications.

When an alloy liner material is to be used, consideration might be given to fabricating the entire duct section from the material specified for the liner. This will generally be more expensive, but, depending on the construction circumstances, may be an attractive alternative.

Internal gutters, downspouts, and other liquid flow collection devices are sometimes used in ducts where the flue-gas temperature is below its acid dew point and excessive internal liquid condensation is expected. When this is the case, internal drains should be installed in the ducts to take the flow to a proper repository. Because of the corrosive nature of this liquid, these liquid flow collection devices and drains should be made from the same material as the duct liner.

Hot gas ducts for gas-turbine exhaust applications use a thin "fish scale" type lining over internal duct insulation, which protects the structural outer shell plate from the high temperatures.

3.8 HANGER ELEMENTS

Ductwork is frequently supported by hangers comprised of hanger rods, clevises, and turnbuckles. Where these hanger elements are exposed to temperatures of less than 750 °F (400 °C), carbon steel rods such as ASTM A36 are typically used with forged steel clevises and turnbuckles. For higher temperatures of up to 1,100 °F (600 °C), alloy steel rods, clevises, and turnbuckles conforming to ASTM A182 Grade F11 are typically used.

References

AISC. 2003. *Fire resistance of structural steel framing: AISC design guide 19.* Chicago: AISC.
AISC. 2017. *Specification for structural steel buildings.* AISC 360-16. Chicago: AISC.

ASME. 2013. *Reinforced thermoset plastic corrosion resistant equipment.* ASME RTP-1. New York: ASME.

ASME. 2015a. *ASME boiler and pressure vessel code, Section II: Materials.* New York: ASME.

ASME. 2015b. *ASME boiler and pressure vessel code, Section IX: Welding, brazing, and fusing qualifications.* New York: ASME.

ASTM. 2011. *Standard test method for pitting and crevice corrosion resistance of stainless steels and related alloys by use of ferric chloride solution.* ASTM G48. West Conshohocken, PA: ASTM.

AWS (American Welding Society). 1996. *Recommended practice for welding of chromium-molybdenum steel piping and tubing.* AWS D10.8. Miami, FL: AWS.

AWS. 2015. *Structural welding code, steel.* AWS D1.1. Miami, FL: AWS.

CEN (European Committee for Standardization). 2013. *Free-standing chimneys, Part 7: Product specifications of cylindrical steel fabrications for use in single wall steel chimneys and steel liners.* EN 13084-7, CEN/TC 250. Brussels, Belgium: CEN.

NACE (National Association of Corrosion Engineers). 2012. *Installation of thin metallic wallpaper lining in air pollution control and other process equipment: Publication SP0292-2012.* Houston, TX: NACE.

CHAPTER 4

Service Conditions and Design Loads

4.1 SERVICE CONDITIONS

Before beginning the structural analysis and design of ductwork and its support structure, the structural engineer should obtain the following pertinent information regarding the service conditions to which the ductwork will be subjected.

4.1.1 Site Location

The structural engineer needs the geographical location of the ductwork system including information about the local topography. This information will be used to determine the applicable wind loads and seismic design category and to ascertain which design codes are applicable, including any local rules and regulations.

4.1.2 Indoor versus Outdoor Service

The structural engineer needs to know whether the ductwork is indoors or outdoors. Indoors, wind and snow loads may be ignored except when addressing construction loading conditions. Lagging requirements may also change if the ducts are indoors.

4.1.3 Mechanical Performance Considerations

Before arranging and designing the supports for the ductwork, the structural engineer should consult with the responsible mechanical process engineers on the project. The structural engineer should obtain all the available information regarding the desired performance of the ductwork system, such as

- Flow rate;
- Size, weight, location, number, and type of flow distribution devices;
- Operating pressure and temperature;
- Transient pressure and temperature;
- Ash depth, density, and locations;

- Type, location, and weight of dampers; and
- Expansion joint types and locations.

4.1.4 Nature of the Media Flowing through Ducts

Information regarding the media flowing through the ducts should be obtained by the structural engineer. For air ducts, it is important to know whether contamination with ash via carryover through common equipment, such as an air preheater, is expected. For flue-gas ducts, it is important to know the chemical composition of the gas and any injected chemicals, such as trona, ammonia, activated carbon, or other pollution control chemicals. It may be important to know whether harmful chemicals, such as chlorides, fluorides, ammonia, and sulfides, are present in the flue-gas. It is also important to know the expected dew point and moisture content of the flue-gas. These data can be used to determine whether a sludge or slurry will form at the expected flue-gas service temperatures and whether special insulation and/or corrosion-resistant materials should be specified at expected cold spots, such as instrument penetrations, access doors, and low-flow regions among others. The expected ash quantity and bulk density in the flue-gas, as well as the velocity, will also help determine what ash loading should be used for the design.

4.1.5 Operating Conditions

The maximum expected continuous operating conditions inside the ductwork should be obtained from the mechanical process engineer or other appropriate sources. The values provided may be somewhat higher than what is commonly called the normal operating conditions, because there may be fluctuations during operation. However, they can persist for an extended period and thus should be the governing criteria.

The operating conditions that are necessary to know are the *maximum expected operating pressure* and the *maximum expected operating temperature*. Because of the ductwork configuration and the equipment in the duct system, the pressure and temperature are often different in different sections of duct. Figure 1-6 shows a typical pressure profile for a ductwork system. For typical *example* pressure and temperature values, see the following tables:

- Pressurized coal-fired power plant, Table 4-1,
- Balanced draft coal-fired power plant, Table 4-2, and
- Pressurized and balanced draft industrial boiler systems, Table 4-3.

4.1.6 Excursion and Transient Conditions

The maximum expected excursion and transient conditions inside the ductwork should also be obtained from the mechanical process engineer or other appropriate sources. These conditions may occur at some point in the operation and exist only for a relatively short time. Depending on the event, this could be seconds, minutes, or maybe even hours.

Table 4-1. Air and Gas Duct Example Pressures and Temperatures for a Pressurized Coal-Fired Power Plant.

Duct section	*Operating pressure (in. H_2O)*	*Transient pressure (in. H_2O)*	*Operating temp. (°F)*	*Excursion temp. (°F)*
Forced draft fan to air preheater	+30 to +60	+60 to +90	50 to 120	100 to 150
Air preheater to boiler	+25 to +55	+60 to +90	300 to 700	400 to 750
Air preheater to pulverizer	+30 to +60	+70 to +90	250 to 700	400 to 750
SCR	+10 to +25	+30 to +40	650 to 850	750 to 950
Air preheater inlet (gas side)	+10 to +25	+30 to +40	650 to 850	750 to 950
Air preheater to precipitator or baghouse	+5 to +20	+30 to +40	250 to 450	350 to 750
Precipitator or baghouse to stack	+5 to +10	+10 to +15, 0 to −10	250 to 450	350 to 750

Notes:

1. The location and closing time of dampers can greatly affect the pressure, especially the transient pressures given in this table. This table assumes that there are dampers at the precipitator inlet.
2. The values in this table are only representative. They should not be used for design. The user should obtain the proper design values for the duct system under consideration from the appropriate sources.
3. The table is based on a cold-side precipitator or baghouse.

Table 4-2. Air and Gas Duct Example Pressures and Temperatures for a Balanced Draft Coal-Fired Power Plant with FGD.

Duct section	*Operating pressure (in. H_2O)*	*Transient pressure (in. H_2O)*	*Operating temp. (°F)*	*Excursion temp. (°F)*
Forced draft fan to air preheater	+20 to +35	+25 to +35	50 to 120	100 to 150
Air preheater to boiler	+15 to +35	+25 to +35, −5 to −15	300 to 700	400 to 750
Air preheater to pulverizer	+40 to +70	+70 to +90	250 to 700	400 to 750
Boiler to SCR	+0 to +20, −20 to −30	+30 to +40, −30 to −40	650 to 850	750 to 950
SCR to air preheater inlet, gas side	+0 to −25	+35 to −50	650 to 850	750 to 950
Air preheater to precipitator or baghouse	+0 to −35	+35 to −50	250 to 450	350 to 750
Precipitator or baghouse to Induced draft (ID) fan	+0 to −50	+35 to −50	250 to 450	350 to 750
ID fan to scrubber (FGD)	+5 to +15	+30 to +45, −5 to −15	250 to 450	350 to 750
Scrubber (FGD) to stack	+5 to +15	+10 to +25, −5 to −15	120 to 350	180 to 750

Notes:

1. The location and closing time of dampers can greatly affect the pressures, especially the transient pressures given in this table. This table assumes that there are dampers at the precipitator inlet.
2. The values in this table are only representative examples. They should not be used for design. The user should obtain the proper design values for the duct system under consideration from the appropriate sources.
3. The table is based on a "cold side" precipitator or baghouse.
4. The table is based upon the balanced point being located within the boiler.
5. The table assumes no ID booster fan is provided. Between the ID fan and the ID booster fan the operating pressures may fluctuate between positive and negative.

Table 4-3. Air and Gas Duct Example Pressures and Temperatures for Pressurized and Balanced Draft Industrial Boilers.

Pressurized System				
Duct section	*Operating pressure (in. H_2O)*	*Transient pressure (in. H_2O)*	*Operating temp. (°F)*	*Excursion temp. (°F)*
FD fan to air preheater	+15 to +30	+20 to +40	50 to 120	100 to 150
Air preheater to boiler	+10 to +25	+15 to +35	300 to 500	400 to 500
Boiler to heat trap (exchanger)	+5 to +20	+10 to +25	450 to 800	500 to 850
Heat trap to air pollution control device (APCD)	+5 to +20	+10 to +25	300 to 500	400 to 600
APCD to stack	0 to +5	0 to +10	250 to 500	300 to 600

Balanced Draft System				
Duct section	*Operating pressure (in. H_2O)*	*Transient pressure (in. H_2O)*	*Operating temp. (°F)*	*Excursion temp. (°F)*
FD fan to air preheater	+10 to +20	+15 to +30	50 to 120	100 to 150
Air preheater to boiler	+5 to +15	+10 to +20	300 to 500	400 to 500
Boiler to heat trap (Exchanger)	−5 to −10	−10 to −15	450 to 800	500 to 850
Heat trap to APCD	−10 to −20	−15 to −20	300 to 500	400 to 550
APCD to ID gan	−10 to −20	−15 to −25	300 to 500	400 to 550
APCD to stack	0 to +5	0 to +5	300 to 500	400 to 550

Notes:
1. The location and closing time of dampers can greatly affect the pressure, especially the transient pressures given in this table. This table assumes that there are dampers at the air pollution control device (APCD) inlet.
2. The values in these tables are only representative. They should not be used for design. The user should obtain the proper design values for the duct system under consideration from the appropriate sources.

Short-term conditions that are necessary to know are the *maximum expected transient pressure* and the *maximum expected excursion temperature*. As with the operating conditions, short-term values will usually be different in different sections of duct. The minimum design pressure requirement given in Chapters 4 and 6 of the National Fire Protection Association's NFPA 85, *Boiler and Combustion Systems Hazards Code* (NFPA 2015) may be applicable. Pressures may be negative or positive. These transient pressures may occur during boiler implosions, flame basket collapse, main fuel trips, failure of dampers to open or close, accidental damper actuation, air preheater failure, or emergency shutdowns.

See Tables 4-1, 4-2, and 4-3 for typical example values of transient pressures and excursion temperatures.

4.1.7 Insulation and Lagging Requirements

The structural engineer must be able to determine or conservatively estimate the weight of the insulation and lagging that will ultimately be attached to the ductwork. The insulation may be inside or outside the duct. Refractory may be used as internal insulation and for corrosion protection. The thickness, type of insulation, and location of any air gaps are often based on the customer's specification and personnel protection, as required, per ASTM C1055 (ASTM 2014), *Standard Guide for Heating System Surface Conditions That Produce Contact Burn Injuries*, which specifies a maximum temperature of 140 °F (60 °C) on the outer surface to protect plant workers. The thickness and type of lagging are also often specified by the customer or may be dictated by the necessity to walk on it. In any event, the structural engineer needs to know enough to be able to obtain a weight for these items, and it is also important to know the thickness and arrangement of the insulation and lagging when calculating area for wind load (Figure 4-1). It is acceptable to make conservative estimates of these weights except with spring-supported systems or where uplift loading is critical. In these cases, a more accurate weight determination is essential for proper load analysis. More information on insulation and lagging can be found in Chapter 11.

4.1.8 Corrosion and Erosion Allowances

The customer may specify a desired corrosion or erosion allowance for the duct plate and internal element. If no allowance is specified, the structural engineer may wish to add one if certain severe conditions exist. Usually, this is in the form of a thickness allowance. See Section 4.1.4 for discussion of the possible damaging elements carried in ducts. Special coatings, linings, and materials can also be used in lieu of a thickness allowance in severe conditions, as described in Chapter 3. If significant, the added weight of these should be considered in the design. In no circumstances should a corrosion or erosion allowance be considered as part of the structure for strength calculations. If a corrosion or erosion allowance is used, the structural engineer should calculate the strength based on the reduced thickness of plates or shapes, but dead loading should be based on full thickness.

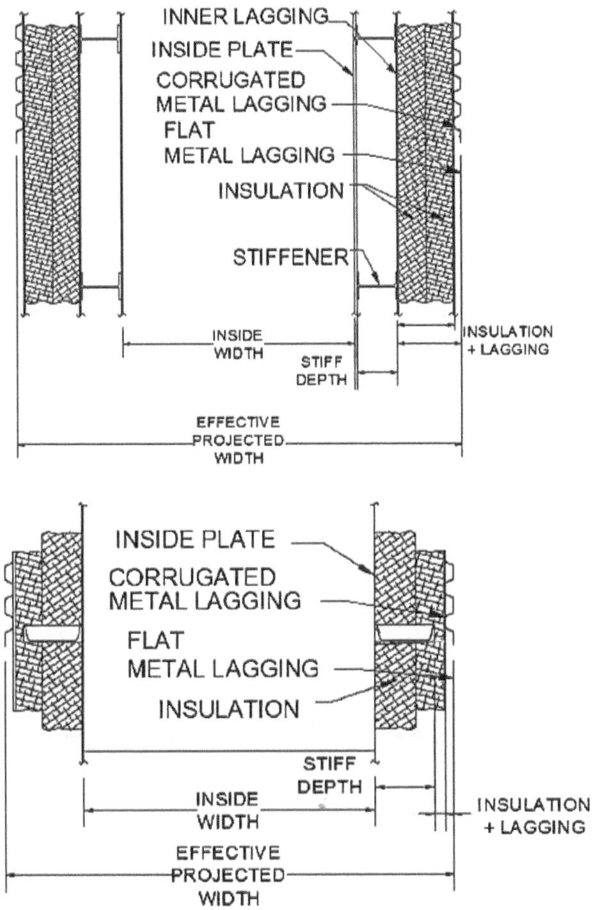

Figure 4-1. Effective wind area.

4.2 DESIGN LOADS

The ductwork should be designed for the following minimum loads and combinations thereof. Loading combinations and appropriate design strengths are presented in Chapter 5.

4.2.1 Dead Load

The dead load to be used in the structural analysis should consist of the weight of all duct components and any other materials permanently fastened thereto.

Dead load includes duct plate, stiffeners, trusswork, turning vanes, platforms, stairs, ladders, dampers, hoppers, expansion joints, subgirts, and any other steel permanently attached. It may also include insulation, refractories, lagging, linings, test ports, instrumentation, or any other equipment supported by or within the duct.

4.2.2 Live Load

Live loads are those produced by the occupancy or use of the ductwork for temporary laydown or storage. Live loads may be uniformly distributed, like a platform supported from the ductwork, or they may be concentrated, such as a person walking on top of the duct, safety tie-off points, a davit, or a jib crane. A minimum design live load of 20 lb/ft^2 (0.96 kPa) is typically used in practice. Whenever the top of a duct may be used for equipment laydown or storage, the duct should be designed for this extra load if it exceeds the minimum design live load. Platform loads are typically a minimum of 75 lb/ft^2 (3.6 kPa). This may be greater if heavier equipment is expected to be transported across the platform. The minimum concentrated load for a man walking on a duct is usually 300 lb (1,334 N).

4.2.3 Environmental Loads

Environmental loads are typically defined by local and state building codes or as presented in ASCE 7-16, *Minimum Design Loads and Associated Criteria for Building and Other Structures* (2017). They are wind load (*W*), ice load (*D*i), snow load (*S*), and earthquake load (*E*).

Ductwork is ordinarily considered by structural engineers a Risk Category III (or IV) structure for wind, snow, and seismic loads, as defined in ASCE 7-16. Wind and snow loads should be based on the exterior dimensions of the ductwork, which should consider the projection of the insulation and lagging beyond the inside face of the duct. This is discussed in Section 4.1.7 and illustrated in Figure 4-1. Snow loads may not be applicable while the duct is operating, because (depending on the project location) there may be enough heat loss through the insulation and lagging to melt the snow. The wind load analysis needs to consider the global main wind-force-resisting system loads for the overall system design, including the trusses and supports. Local component and cladding wind loads are used for the lagging and local stiffener design. Wind loads are not normally considered in the local duct plate design, because the insulation and lagging subgirts typically span between stiffeners, so no wind is applied to the duct plate. Seismic loads, determined in accordance with ASCE 7-16 (Sections 11, 12, 13, and 15), are usually based on dead load, full design ash loads, and a portion of the snow load where the flat roof snow load exceeds 30 lb/ft^2 ft (1.44 kPa).

If the ductwork is indoors, wind and snow loads may be neglected except when designing for construction loading conditions. If the ducts are not designed for wind or snow loads, the structural engineer should advise the owner and the erector, especially if snow or wind loads are expected to be significant (cause damage) during construction. See Section 4.2.8 for more on construction loads.

Ice loads are not typically considered in design because in most cases ductwork is not considered an ice-sensitive structure, and the ice may be melted when the duct is in operation (hot). Ice loads should be considered if applicable based on the ductwork arrangement and project location, and the load combinations should be modified to include ice loads based on the recommendations in ASCE 7-16.

4.2.4 Pressure Loads

Pressure loads are loads resulting from the operation of the ductwork system. These loads may result from steady-state or short-term conditions.

Expansion joints are added to the ductwork system to absorb and control thermal expansion. Whenever a section of duct has an expansion joint located so that there is a duct wall that does not have another duct wall directly across from it, an *unbalanced pressure* occurs in that duct section. Unbalanced pressure forces will only occur opposite openings or expansion joints upstream or downstream of bends or transitions. When this arrangement occurs, the resulting unbalanced pressure force and any associated moments must be transferred through the duct to its support structure. The line of action of this force is along the centerline axis of the expansion joint perpendicular to the opening. The magnitude of the force is calculated by multiplying the area of the opening by the internal pressure, either operating or transient, depending on the condition being considered. In actuality, the flow in ducts obeys Bernoulli's equations for mass flow of a compressible fluid; but because the mass velocity term is small compared to the pressure and area terms, it can usually be ignored. Figure 4-2 shows an example of the effects on a duct section of unbalanced pressure. For detailed discussion of unbalanced pressure, see Section 2.7.

Operating pressure is the expected maximum operating air or flue-gas pressure continuously acting on the walls and flow distribution devices in the ductwork over an extended period.

Transient pressure is the maximum expected air or flue-gas pressure acting on the walls and flow distribution devices in the ductwork for a relatively short duration: seconds, minutes, or hours. Transient pressure is usually considered as an accident or extraordinary-event condition.

4.2.5 Ash Loads

When fly ash is entrained in flue-gas, ash fallout onto the duct floor should be expected in areas of low flow velocity. Thus, ash loading should be considered in the design of the flue-gas ductwork. Figure 4-3 shows typical locations of ash accumulation. Note that there may be ash deposits on the duct floor downstream of a precipitator or fabric filter. Air ducts can be contaminated with ash via carryover through equipment such as an air preheater but ash loads are typically not considered in air ducts with a tubular air preheater.

Because ash loading can be significant, affecting the cost of the ductwork and its supporting structure, flow model studies are often used to predict locations of ash fallout. These studies can be costly and time-consuming, and may not quantify the magnitude of ash accumulation, so the engineer and the owner should weigh the benefits of the flow model against their cost and overall schedule requirements. Flow models, either physical or computational fluid dynamic (CFD), are also useful in determining flow distribution and pressure drop and often are more accurate in the determination of these effects than analytical methods. All this should be considered in the decision process.

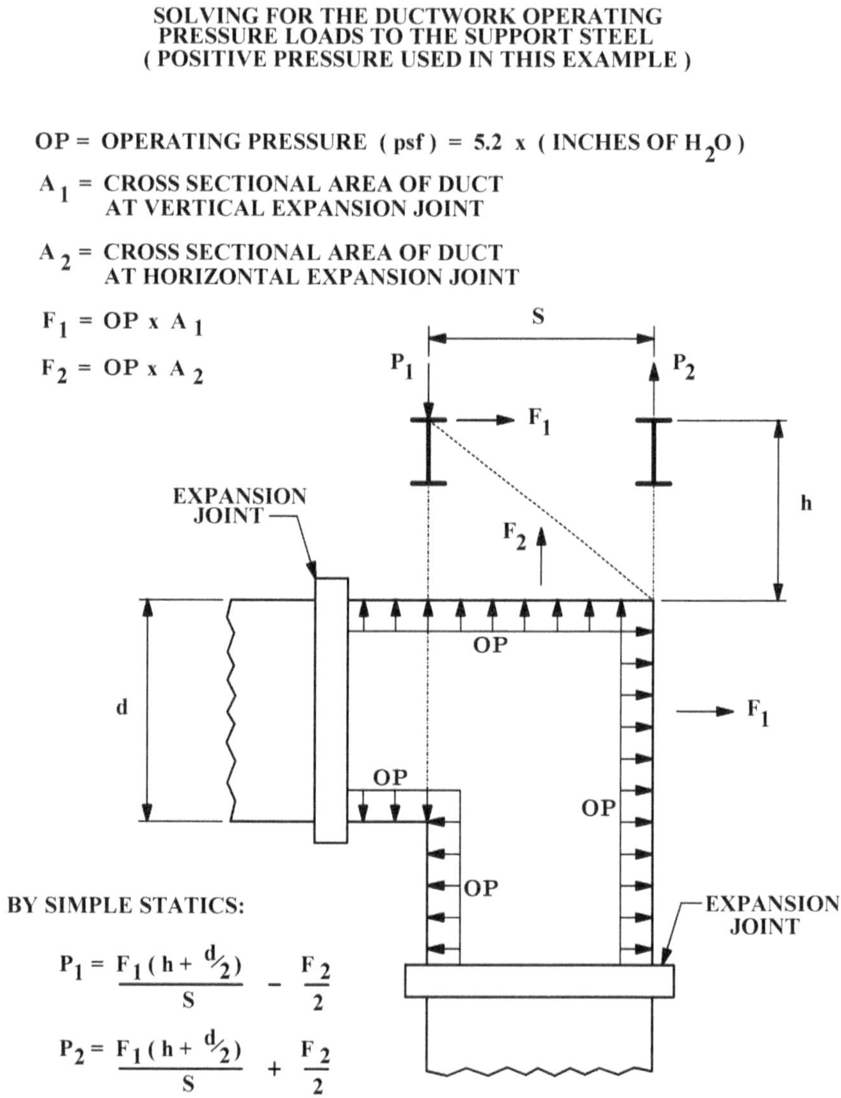

Figure 4-2. Effect of unbalanced pressure forces on ductwork and its external supports.

Ash densities depend on the type of fuel burned, the flue-gas temperature and humidity, the method of combustion, and the wetness of the ash. A minimum pulverized coal ash density of 75 lb/ft^3 (1,200 kg/m^3) is recommended, but the use of a higher value may be necessary, depending on the exact type of coal or lignite and the plant operating conditions. Different values may be used for the ash density from burning oil, wood, bark, bagasse, black liquor, and municipal waste fuels (Table 4-4).

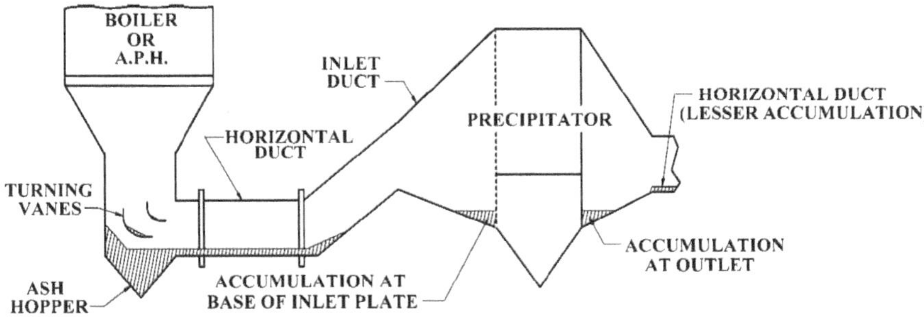

Figure 4-3. Typical locations of ash accumulation.

Table 4-4. Fly Ash Density Ranges.

	Density	
Fuel	*(lb/ft³)*	*(kg/m³)*
Coal or lignite	75–120	1,200–1,922
Oil	10–40	160–641
Wood, dry	10–30	160–481
Bark, bimodal sand	100–120	1,602–1,922
Bark, char	10–30	160–481
Bark, combination	40–80	641–1,281
Bagasse	30–50	481–801
Black liquor	10–20	160–320
Municipal waste	30–60	481–961

Special consideration is needed in ducts with low flow velocities, where large ash fallout may occur. Ash depths can be as high as 5 to 8 ft (1.5 m to 2.4 m) in the worst conditions and large deposits of ash act as a thermal insulator that can introduce thermal variations in stiffeners and casing, as discussed in Section 4.2.7. Typical design depths vary from 6 in. (152 mm) to 25% of the duct height. The amount of ash to be considered in the analysis and design of the ducts depends on several factors including the following.

Type of fuel burned. Ash is more predominant in units fired by coal, black liquor, and solid fuel than in oil-fired or gas-fired units, where it is often neglected.

Flow velocity. Fly ash is likely to settle out in ducts where the velocity is less than 40 ft/s (12.2 m/s). Bypass or crossover ducts that have no flow through them in many modes of operation are particularly prone to accumulating large amounts of ash.

Location relative to the ash collection equipment. Ducts upstream of ash collection equipment such as mechanical dust collectors, precipitators, baghouses, and scrubbers will accumulate more ash than ducts downstream of the equipment. But very large accumulations of ash are also common on the floor of the duct at the outlet of the ash collection equipment. Flow distribution baffles placed at the inlet of ash collection equipment often catch ash and have a fairly large accumulation at their base.

Increase in cross-sectional area. When the cross-sectional area of a duct greatly increases, the velocity greatly decreases, and there is a strong likelihood that large amounts of ash will fall out onto the duct floor.

Changes in flow direction (duct corners). Ash is more apt to settle out where the ductwork abruptly changes direction. There is a localized increase in the cross-sectional area at duct bends, which in turn reduces the flow velocity. There may also be some flow stratification, leading to localized areas of slower flow, which in turn increases the amount of ash fallout. One example of this is where a horizontal duct takes a 90-degree turn. Other examples are where ductwork goes from vertically down to horizontal or horizontal to vertically up. At these locations, hoppers may be provided. They should be designed to be completely full of ash to the bend line. In situations where the reliability of the ash removal system is suspect, it may be prudent to design for an even higher level.

Long horizontal or slightly inclined ducts. These ducts are also susceptible to ash accumulation. Project design criteria should establish a minimum design depth based on anticipated operating conditions. Typically, ducts that are inclined more than 45 degrees from the horizontal accumulate a negligible amount of fly ash, because most of the ash fallout slides down to an adjacent horizontal surface.

Turning vanes and flow devices. Some turning vanes act as a ledge and tend to collect ash. Other vanes may have small spacing and tend to collect ash. They should be designed accordingly. See Chapter 9 for more information on turning vane loadings.

Cycling units. Cycling units can have low flow rates during which ash is deposited, followed by high flow rates that wash it away. Units operating at significantly reduced load produce less fly ash, because less coal is being burned per hour and typically a larger percentage of the heavier ash particles are retained in the boiler as bottom ash, because flow velocities are not sufficient to entrain them and carry the particles into adjacent ducts.

Dampers. Stacked dampers with the bottom damper closed will cause ash accumulation.

Riser tubes, tube sheet, and other internal obstacles. All obstacles to the flow, such as flow through the economizer, tend to cause flow eddies and/or velocity changes that lead to ash deposition. Boiler tube pipe leakage may also occur, resulting in water mixing with the fly ash, which increases the fly ash density.

Scrubbers (wet and dry). Backwash and/or ash accumulation before or after scrubbers (dry or wet) are possible and need to be considered in the duct design.

Boiler or equipment tube leak water. Water from an equipment failure such as a boiler tube rupture will increase the ash weight in nondrainable areas and should be considered.

Wash water. Water used to clean the duct will increase the weight of the accumulated ash, but this would be during an outage and therefore does not influence the operating loads.

Sludge loads. Sludge loads are similar in nature and behavior to ash loads. Downstream of wet and dry scrubbers and condensing heat exchangers, moisture can result as the flue-gas cools to below its dew point. The carry-over from this process often results in sludge on the interior surfaces of the duct. Where coal sludge deposits are expected, a minimum density of 125 lb/ft^3 (2,000 kg/m^3) is typical, but the use of a higher value may be necessary, depending on the type of fuel, the flue-gas moisture content, and other operating conditions. It is not uncommon to find 6 in. to 12 in. (152 mm to 305 mm) of sludge buildup on the walls and ceiling of scrubber outlet ducts. Much larger amounts have been known to accumulate, depending on the frequency of cleaning and removal. A discussion with the owner regarding the cleaning and maintenance of these ducts may be appropriate. More sludge may be present on the duct floor and on turning vanes than on the duct walls.

The load combinations presented in Chapter 5 under the load and resistance factor design method assign the same load factor to ash load as to dead load based on the assumption that the ash load is well defined. If there may be variability in the ash depth or density such that the load magnitude is not considered well defined, then the calculation of the ash load (or the load factor applied to the ash load) should be adjusted accordingly based on the discretion of the engineer, to maintain a safety factor consistent with the methodology presented in ASCE 7-16.

4.2.6 System-Related Loads

System loads (SL) are loads that are created or result from decisions made during the design and layout process. They become real loads on the system.

Dynamic unbalanced pressure forces, SL_{DP}. Whenever turns, elbows, or flow distribution devices, such as turning vanes, splitter plates, or perforated baffle plates, are added to the system to distribute flow or reduce pressure drop, an unbalanced dynamic pressure force is created. The structural engineer should determine whether these forces are small enough to be neglected or large enough that they should be considered in the design. With turning vanes and perforated baffle plates, they should be considered. See Chapter 9 for the procedure to calculate dynamic unbalanced pressure forces and the structural design of flow distribution devices.

Expansion joint actuation forces, SL_{EJ}. For an expansion joint to operate (compress and extend) as intended, a force must be applied to it. If a joint is very flexible, like many nonmetallic expansion joints, this load is very small and may be ignored. If, however, relatively stiff metal joints are used, this force may be large and need to be considered when resolving loads to the external support system. The spring constant of the joint is usually available from the joint manufacturer, and this stiffness factor k, along with the anticipated movement, can be used to determine the force that results. If the duct sections undergo rotation along with axial movement, a global moment will be exerted on the duct sections. Formulas for determining these forces and moments are given in the standards of the Expansion Joint Manufacturers Association.

Spring loads, SL_S. If, because of anticipated thermal movement, the structural engineer needs to use spring supports at a certain location, these springs should be treated as loads when resolving the forces to the remaining duct supports for all loading combinations. When analyzing a duct section with spring supports, reverse loads that tug on the duct should be applied at the spring locations. The springs should not be modeled as supports.

There is only one cold setting and one hot setting to which a variable spring can be set. In the case of a constant-support spring, there is also an inertial force of between 3% and 5% of the set load required to drive the spring, which should be considered.

Friction forces, SL_F. When a duct section is base supported, a sliding detail to facilitate thermal growth is required at all base supports except at the duct section's anchor or fixed point. The sliding may take place between two steel surfaces or between one steel surface and a reduced-friction slide bearing plate or between two reduced-friction plates. Slide bearing plates are relatively inexpensive and are frequently used. Rollers are also used in some applications. The anticipated frictional forces should be taken into the duct system and resolved to the fixed supports. The coefficient of friction that is assumed in the structural analysis should be based on the expected value toward the end of the plate's design life, not the new plate's published value. It is not uncommon to use a coefficient of friction of between 0.05 and 0.15 for new designs.

Also, if guide bars, lateral bumpers, or other means are used to transfer lateral loads from wind, seismic, unbalanced pressure, and other loads to the support system, then the lateral support may be engaged during thermal expansion or contraction. Friction forces parallel with the guide bars or lateral bumpers can occur and should be considered in the design at the surface where bearing occurs. The kinetic coefficient of friction is typically used in the design.

4.2.7 Thermal Gradient Loads

The air or flue-gas temperature in any section of ductwork is not always uniform across its cross section. This leads to temperature differentials across the ductwork structure. Also, some of the ductwork elements may be exposed to hot air or gas on one side and ambient air on another side, even if insulated. This leads to a temperature differential across the individual element as well as somewhat of an overall temperature differential across the ductwork structure. Local failures of the insulation system accentuate these conditions.

The deposition of fly ash in ductwork can also create thermal gradient loads (TL). This is because the ash acts as an insulator. Therefore, during unit start-up a condition can occur when the top section of a duct is hot while the bottom is cold, creating a thermal gradient load in the ducts. Of course, the reverse gradient could occur during a unit shutdown. Temperature differentials across an individual member or across a complex structural system can also cause secondary thermal stresses. If the temperature differential is very high, these stresses can be very high, and local failures can result.

Usually, these types of thermal loads are ignored when analyzing and designing ducts. This is because the temperature differentials and their associated stresses are usually very small, the duct structural system is usually flexible enough so that the secondary stresses are easily relieved, and most often temperature differentials are not anticipated. If significant temperature differentials are expected, they should be considered in the structural analysis and design.

The duct engineer should also ensure that appropriate insulation and lagging arrangements are used to limit secondary stresses, such as at large stiffeners or at braced support legs that may not be very flexible.

4.2.8 Construction Loads

It is the erector's responsibility to secure the ductwork against all expected and unexpected loadings during construction until the ductwork is fully assembled and supported. This is in accordance with the AISC *Code of Standard Practice* (AISC 2016). This includes the determination, supply, and installation of shipping bracing, lifting lugs, temporary supports, temporary erection bracing, temporary guys, falsework, cribbing, and anything else required. The erector should be strongly encouraged to consult with the ductwork structural design engineer if he or she has any doubts as to how this is to be accomplished. This practice especially applies to knock-down construction.

However, when ductwork is to be supplied to the field in large shop-assembled sections, the structural engineer should be more involved in this process, as discussed in Chapter 10. In these instances, the structural engineer should select lift points and may design lifting lugs and frames. A lifting diagram may also be required, showing how the piece is to be handled during various phases as it travels from the fabrication shop to its final installation position. The design should consider the mode of transportation. The structural engineer may have to consult with the fabricator and erector as to the lifting equipment to be used. When designing these lifting lugs or frames, an impact factor of 1.5 should be applied to the lifted loads. Lifting loads should also be given to the structural steel design engineer. The number of lifting points is often smaller than the number of final support points, so structural steel in the area should be designed accordingly. More guidance is provided in ASME BTH-1-2017, *Design of Below-the-Hook Lifting Devices* (ASME 2017).

Circular ducts have a special sensitivity to transportation and handling when they are shipped in one piece as a cylinder, because of their potential for buckling, as explained in Chapter 7. Analyzing and designing these ducts for impact during shipping and handling should be considered. See Chapter 10 for more details on bracing for shipping and all the other construction-related topics mentioned in this section.

REFERENCES

AISC. 2016. *Code of standard practice for steel buildings and bridges*. AISC 303. Chicago: AISC.

ASCE. 2017. *Minimum design loads and associated criteria for building and other structures*. ASCE 7–16. New York: ASCE.

ASME. 2017. *Design of below-the-hook lifting devices: BTH-1-2017*. Washington, DC: ASME.

ASTM International. 2014. *Standard guide for heated system surface conditions that produce contact burn injuries*. ASTM C-1055. West Conshohocken, PA: ASTM.

Expansion Joint Manufacturers Association. 1993. *Standards of the expansion joint manufacturers association*, 10th ed. Terrytown, NY: Expansion Joint Manufacturers Association.

NFPA (National Fire Protection Association). 2015. *Boiler and combustion systems hazards code: NFPA 85*. Quincy, MA: NFPA.

CHAPTER 5

Load Combinations and Associated Design Strengths

5.1 DESIGN CONSIDERATIONS

The structural design of air and gas ducts involves loads that are unique to these structures, as well as loads that are encountered in the design of any typical structure. These loads must be combined to determine the required strengths of the various structural members. The loads are discussed in detail in Chapter 4, and load combinations are discussed here.

Some of the loads occur at their full value over extended periods of time and are considered *sustained* loads. Other loads occur at their full value only over short periods of time and are considered *transient* loads. Also, the ductwork is exposed to different temperatures during its lifetime, and not all loads will occur at all temperature conditions. Each of these situations should be considered in the structural design. The structural engineer must design ductwork that will be reliable, operate safely, and not fail under all the anticipated conditions for the design life of the duct. This section will provide guidance for doing so.

5.2 STRESS-BASED VERSUS STRENGTH-BASED DESIGN

The structural design of air and gas ducts is traditionally done on the basis of AISC's *Specification for Structural Steel Buildings* (AISC 2017a). Although this specification is intended for the design of steel buildings and other structures that are not typically exposed to elevated temperatures, it is relatively straightforward to adapt the equations therein for designing at elevated temperatures. The accepted practice in the ductwork industry has been to use these equations with temperature-adjusted values for the yield stress F_y, tensile strength F_u, and modulus of elasticity E obtained from testing performed on steel at elevated temperatures. When the temperature becomes high enough that creep becomes a concern, additional considerations should be taken into account. Creep was discussed in detail in Chapter 3, and creep in structural design is further discussed in Section 5.3.

Through the 1989 version, the AISC specification used an allowable-stress design approach. However, the more recent editions have changed to strength-based design. Thus, to use the current specification (*Specification for Structural Steel Buildings*, ANSI/AISC 360-16) (AISC 2017a) to design the structural components of ductwork, a shift to a strength-based design approach is required. The specification permits design according to one of two methods: allowable-strength design (ASD) and load and resistance factor design (LRFD). In ASD, the required strength of a member is determined using load combinations consisting of service-level loads. This required strength is compared to the allowable strength of the member, which is the nominal strength, R_n, divided by a factor of safety, Ω. The member is adequate if the required strength is less than or equal to the allowable strength. In LRFD, the required strength of a member is determined using load combinations in which the loads are factored up to an ultimate strength level. This required strength is compared to the design strength of the member, which is the nominal strength R_n multiplied by a resistance factor ϕ to account for unavoidable deviations of the actual strength from the nominal strength. The member is adequate if the required strength is less than or equal to the design strength. Note that both methods make use of the same nominal strength, R_n.

5.3 STRENGTH AT TEMPERATURES WITHIN THE CREEP RANGE

As discussed in Chapter 3, creep is a stress-related issue that is time-dependent. At high enough temperatures, the steel will begin to creep, where the strain increases over time even though the stress remains constant. Given enough time, failure can occur at stress levels well below the yield stress. Therefore, when the design temperature is within the creep range, creep becomes a primary factor in determining the strength of the steel members.

The creep strength decreases not only with increasing temperature but also with longer exposure to load. A member can be allowed to experience a higher stress if it only must resist that stress for a short period than if it has to resist it for longer. It is therefore important to select the appropriate exposure time for a given load when determining the limiting creep stress that will be used to design the members for that load.

The approach suggested herein for when operating temperatures are within the creep range is to use two sets of load combinations: one that includes only long-sustained loads, and another involving short-duration loads, such as full live, snow, wind, earthquake, and transient pressure loads. Each set of load combinations links to a different limiting creep stress. For the long-duration load combinations, it is suggested that the limiting creep stress be based on 100,000 h data. For the short-duration load combinations, it is suggested that the limiting creep stress be based on 100 h data unless longer load durations are expected. Refer to Chapter 3 for guidance on determining these stresses.

For design, the limiting creep stress becomes the basis for calculating the nominal creep strength of the member, R_n. By applying a factor of safety to the nominal creep strength (ASD method) or factoring the service level loads up to an ultimate level plus applying a resistance factor to the nominal creep strength (LRFD method), the actual expected time-to-rupture under the service-level loads will be much greater than the limiting creep stress basis and well in excess of the actual total load duration during the expected life of the ductwork.

It should be noted that the actual expected time-to-rupture under service-level loads will not necessarily be the same with the LRFD method as with the ASD method. This is because, in the ASD method, the factor of safety Ω is constant for a given limit state, whereas in the LRFD method, the equivalent factor of safety is not constant. In LRFD, the load factor for sustained or permanent loads is generally lower than the load factor for transient loads. Thus, the factor of safety depends on the average load factor, which in turn depends on the ratio of sustained loads to transient loads in a given load combination, and the resistance factor ϕ that is applied to the nominal creep strength. If the average load factor divided by the resistance factor is less than the factor of safety used in ASD, the actual expected time-to-rupture will be less with the LRFD approach than with the ASD approach if the same nominal creep strength R_n is used.

5.4 LOAD DEFINITIONS

The design loads listed below are defined and explained in Section 4.2. The symbols used in the load combinations follow:

A	Ash load (or sludge load)
CL	Construction loads
D	Dead load
E	Seismic load
L	Live load
OP	Operating pressure
S	Snow load
TL	Thermal gradient loads
TP	Transient pressure
W	Wind load
$\sum$SL	System-related load: summation of system loads that act simultaneously and may include

SL_{DP}	Dynamic unbalanced pressure forces,
SL_{EJ}	Expansion joint actuation forces,
SL_S	Spring loads, and
SL_F	Friction forces.

5.5 DEVELOPING LOAD COMBINATIONS FOR DUCTWORK DESIGN

5.5.1 General Requirements

To properly design the ductwork, the structural engineer should consider the loads outlined in Chapter 4 and evaluate the probability of their simultaneous occurrence. The structural engineer should also determine whether the loads are sustained or short-term and the maximum temperature under which they are likely to exist. Load combinations should be developed that consider each of these factors. The required strengths resulting from the selected load combinations should then be compared to the appropriate limiting strength.

The current standard used to develop load combinations is ASCE 7-16, *Minimum Design Loads and Associated Criteria for Buildings and Other Structures* (ASCE 2017). The load combinations to be used for ductwork design should be based on the basic load combinations found in ASCE 7, expanded to include the additional loads that are unique to ductwork design. When doing so, the basic philosophy behind the ASCE 7 load combinations should be maintained. This basic philosophy is that, except for permanent loads such as dead load, most loads are transient: they vary with time. When any one transient load is at its full value, the other transient loads are either at a reduced arbitrary point-in-time value or are not present at all. The maximum values of two or more transient loads are never combined in one load combination. To expand on this philosophy, the unique ductwork loads should be classified as either sustained or transient. Sustained loads are loads that will always be present during normal operation and include dead load, ash load, operating pressure load, and system-related loads (operating values). These loads should be included at their maximum values in all load combinations, except for opposing-force combinations in some cases. See further discussion in Section 5.5.5 for how to handle pressure load, ash load, and system-related loads in opposing-force load combinations.

5.5.2 LRFD Load Combinations That Include Loads Not Specified in ASCE 7

For ductwork design, load combinations must be developed that include loads that are not specified in ASCE 7, such as ash load, pressure load, and system-related loads. Section 2.3.5 of ASCE 7-16 gives requirements for developing LRFD load combinations that include nonspecified loads. Section 2.3.5 requires the use of a probability-based method that is consistent with the method used by the ASCE 7 committee to develop the basic LRFD load combinations. Section C2.3.5 of the commentary to ASCE 7-16 provides guidance on using statistical analysis to determine a mean value and coefficient of variation to set load factors for loads that are not directly covered therein.

There is currently not enough collected data on loads such as ash load or pressure load to allow such statistical analyses. Therefore, the nonspecified loads are handled as follows in the LRFD load combinations given in Section 5.7:

- Operating pressure is treated as a fluid load because it is well-defined. According to ASCE 7, fluid load shall have the same load factor as dead load. Therefore, the load factor on operating pressure matches the load factor on dead load (except in opposing-force combinations), and operating pressure is included in all load combinations except those with transient pressure.
- Transient pressure is treated as an extraordinary event because of its low probability of occurrence and short duration. Therefore, it is combined with other loads according to Equation 2.5-1 in Section 2.5.2.1 of ASCE 7-16, with the load factor on transient pressure set to 1.
- Ash load is treated as a sustained load and is assigned the same load factor as dead load in the LRFD load combinations. Because of this, the ash load should account for buildup; see the discussion of ash load in Chapter 4. There may be cases where the structural engineer will want to increase the ash load factor in the LRFD load combinations, such as locations where the ash depth is not well defined and is transient.
- System-related loads are treated as sustained loads and assigned the same load factor as dead load except in opposing-force combinations.

5.5.3 ASD Load Combinations That Include Loads Not Specified in ASCE 7

There are no requirements given in ASCE 7-16 for including nonspecified loads in ASD load combinations. The ASD load combinations given in Section 5.6 handle these loads as follows:

- Operating pressure is treated the same as in LRFD load combinations: as a fluid load. It is included in all load combinations except those with transient pressure.
- Transient pressure is treated as an extraordinary event, just as in LRFD load combinations. ASCE 7-16 does not address ASD load combinations with extraordinary events, so the ASD load combination has been developed from the LRFD load combination by multiplying the load factors in the LRFD combination by 0.67. The 0.67 factor is the ratio of the ASD allowable strength (R_n/Ω) to the LRFD design strength (ϕR_n). It is determined from the relationship between the LRFD resistance factors and the ASD safety factors that was developed by AISC, which is $\Omega = 1.5/\phi$. (Refer to the commentary to Section B3.4 of the AISC specification for further details.)
- Ash load and system-related loads are treated as sustained loads.

5.5.4 Considerations for Load Combinations with Excursion Temperature

Excursion temperature is a special case. Because of its low probability of occurrence and short duration, it is considered an extraordinary event, like transient pressure. However, because it is not a *load* per se, the fact that it is an extraordinary event cannot be accounted for with a reduced factor on any one load effect, as is done with transient pressure. Instead, it is accounted for by calculating the required strength using the load combination given in Equation 2.5-1 in Section 2.5.2.1 of ASCE 7-16 as the basis but only including *sustained* loads. With the LRFD method, the load factors for the sustained loads match the load factor for dead load in Equation 2.5-1, and the resulting required strength is compared to the design strength at the excursion temperature. With the ASD approach, the factors in the LRFD equation are multiplied by 0.67 (based on the same reasoning discussed under transient pressure in Section 5.5.3), and the resulting required strength is compared to the allowable strength at the excursion temperature.

5.5.5 Other Considerations in Developing and Using the Load Combinations

Operating pressure is included in the opposing-force load combinations shown in Sections 5.6 and 5.7. However, the full value of operating pressure may not always be present and using the maximum value could therefore be unconservative where the operating pressure produces forces that resist the uplift forces from other loads. It is therefore suggested that an additional opposing-force combination be included, with operating pressure set to a smaller value or even zero.

Ash load is not included in the opposing-force combinations with wind or transient pressure, because it may not be present to resist uplift forces when those loads occur. Ash load is included in the opposing-force combinations with earthquake, but only the amount of ash load that was used to determine the earthquake forces should be included in the resistance to uplift.

System-related loads are included at their full values in opposing-force combinations.

The load combinations given in Sections 5.6 and 5.7 do not consider that various power plant loads, such as transient pressure and system loads, may be either positive or negative (friction forces due to expansion and contraction, for example). This should be considered in developing the actual load combinations for design. The total number of load combinations will therefore be much higher than the generic combinations given in this book.

The load combinations given in Sections 5.6 and 5.7 include the ASCE 7 load combinations, which consider both live load and snow load acting at the same time. With air and gas ducts, this may not be possible in certain situations, in which case the engineer should consider deleting those combinations.

The load combinations given in Sections 5.6 and 5.7 do not include ice load. If ice load is a consideration in the design of the ductwork, include additional load combinations with ice load D_i, per Section 2.3.3 (LRFD) or Section 2.4.3 (ASD) of ASCE 7.

5.6 LOAD COMBINATIONS FOR ALLOWABLE-STRENGTH DESIGN

The following subsections provide load combinations for use with the ASD method. These load combinations consider the temperature, type of load, and duration of load, and present the recommended allowable strengths that are applicable to the load combinations.

In the load combinations with wind, it is assumed that the wind load has been calculated using a 2010 or later edition of ASCE 7 in which wind has been converted to an ultimate load. Therefore, for ASD, an 0.6 factor has been applied to the wind. If an earlier edition of ASCE 7 is used to calculate wind, the 0.6 factor should be eliminated, and the full value of the wind used.

5.6.1 Load Combinations at Ambient Temperature (Off-Line Condition)

The following ASD load combinations are applicable to off-line conditions at ambient temperature. Allowable strengths at ambient temperature shall apply.

$D + A + L$

$D + A + S$

$D + A + 0.75L + 0.75S$

$D + A + 0.6W$

$D + A + 0.7E$

$D + \mathrm{CL}$

$0.6D + 0.6W$

$0.6(D + A) + 0.7E$

5.6.2 Load Combinations at Operating Temperatures below Creep Range

The following ASD load combinations include sustained plus transient loads at the operating temperature. Allowable strengths, calculated using material properties reduced for the elevated temperature effects, shall apply.

$D + A + \mathrm{OP} + \sum \mathrm{SL} + L$

$D + A + \mathrm{OP} + \sum \mathrm{SL} + S$

$D + A + \mathrm{OP} + \sum \mathrm{SL} + 0.6W$

$D + A + \mathrm{OP} + \sum \mathrm{SL} + 0.7E$

$D + A + \mathrm{OP} + \sum \mathrm{SL} + 0.75L + 0.75S$

$D + A + \mathrm{OP} + \sum \mathrm{SL} + 0.75L + 0.75(0.6W \text{ or } 0.7E)$

$D + A + 0.67\mathrm{TP} + \sum \mathrm{SL} + (0.33L \text{ or } 0.13S)$

$0.6D + \mathrm{OP} + \sum \mathrm{SL} + 0.6W$

$0.6D + 0.6A + \mathrm{OP} + \sum \mathrm{SL} + 0.7E$

$0.6D + 0.67\mathrm{TP} + \sum \mathrm{SL}$

5.6.3 Long-Duration Load Combination at Temperatures within the Creep Range

This ASD load combination includes sustained loads only and is considered a long-duration condition. The resulting required strength should be compared to an allowable strength determined with a limiting creep stress based on 100,000 h load duration, with an appropriate factor of safety.

$D + A + \mathrm{OP} + \sum \mathrm{SL}$

5.6.4 Short-Duration Load Combinations at Temperatures within the Creep Range

The following ASD load combinations include sustained plus transient loads and are considered short-duration conditions. The resulting required strengths should be compared to an allowable strength determined with a limiting creep stress based on an appropriately short cumulative duration, with an appropriate factor of safety.

$D + A + \mathrm{OP} + \sum \mathrm{SL} + L$

$D + A + \mathrm{OP} + \sum \mathrm{SL} + S$

$D + A + \mathrm{OP} + \sum \mathrm{SL} + 0.6W$

$D + A + \mathrm{OP} + \sum \mathrm{SL} + 0.7E$

$D + A + \mathrm{OP} + \sum \mathrm{SL} + 0.75L + 0.75S$

$D + A + \mathrm{OP} + \sum \mathrm{SL} + 0.75L + 0.75(0.6\mathrm{W} \text{ or } 0.7E)$

$D + A + 0.67\mathrm{TP} + \sum \mathrm{SL} + (0.33L \text{ or } 0.13S)$

$0.6D + \mathrm{OP} + \sum \mathrm{SL} + 0.6W$

$0.6D + 0.6A + \mathrm{OP} + \sum \mathrm{SL} + 0.7E$

$0.6D + 0.67\mathrm{TP} + \sum \mathrm{SL}$

5.6.5 Load Combination at Excursion Temperature

This ASD load combination includes sustained loads at the excursion temperature. Allowable strengths, reduced for the temperature effects, shall apply.

$0.8[D + A + \mathrm{OP} + \sum \mathrm{SL}]$

5.6.6 Thermal Gradient Load Combinations at Operating Temperature

The following ASD load combinations include thermal loading conditions at the operating temperature. These load combinations are usually considered only if the structural engineer expects significant thermal loads. Allowable strengths, reduced for the temperature effects, shall apply. If the operating temperature is within the creep range, an allowable strength determined with a limiting creep stress based on 100,000 h load duration, with an appropriate factor of safety, shall apply to the first load combination, and an allowable strength determined with a limiting creep stress based on an appropriately short load duration, with an appropriate factor of safety, shall apply to the second load combination.

$$D + A + \mathrm{OP} + \sum \mathrm{SL} + \mathrm{TL}$$

$$D + A + \mathrm{OP} + \sum \mathrm{SL} + 0.75\mathrm{TL} + 0.75L + 0.75S$$

5.7 LOAD COMBINATIONS FOR LOAD AND RESISTANCE FACTOR DESIGN

The following subsections provide the load combinations for use with the LRFD method. These load combinations consider the temperature, the type of load, and the duration of load, and present the recommended design strengths that are applicable to the load combinations.

In the load combinations with wind, it is assumed that the wind load has been calculated using a 2010 or later edition of ASCE 7, in which wind has been converted to an ultimate load. Therefore, for LRFD design, full wind load has been assigned a load factor of 1. If an earlier edition of ASCE 7 is used to calculate wind, the wind load factors should be multiplied by 1.6.

5.7.1 Load Combinations at Ambient Temperature (Off-Line Condition)

The following LRFD load combinations are applicable to off-line conditions at ambient temperature. Design strengths at ambient temperature shall apply.

$$1.4D + 1.4A$$

$$1.2D + 1.2A + 1.6L + 0.5S$$

$$1.2D + 1.2A + 1.6S + (L \text{ or } 0.5W)$$

$$1.2D + 1.2A + 1.0W + L + 0.5S$$

$$1.2D + 1.2A + E + L + 0.2S$$

$$1.2D + 1.6\mathrm{CL}$$

$$0.9D + 1.0W$$

$$0.9D + 0.9A + E$$

5.7.2 Load Combinations at Operating Temperatures below Creep Range

The following LRFD load combinations include sustained plus transient loads at the operating temperature. Design strengths, reduced for the temperature effects, shall apply.

$1.4D + 1.4A + 1.4\mathrm{OP} + 1.4\sum \mathrm{SL}$

$1.2D + 1.2A + 1.2\mathrm{OP} + 1.2\sum \mathrm{SL} + 1.6L + 0.5S$

$1.2D + 1.2A + 1.2\mathrm{OP} + 1.2\sum \mathrm{SL} + 1.6S + (L \text{ or } 0.5W)$

$1.2D + 1.2A + 1.2\mathrm{OP} + 1.2\sum \mathrm{SL} + 1.0W + L + 0.5S$

$1.2D + 1.2A + 1.2\mathrm{OP} + 1.2\sum \mathrm{SL} + E + L + 0.5S$

$1.2D + 1.2A + \mathrm{TP} + 1.2\sum \mathrm{SL} + 0.5L + 0.2S$

$0.9D + 1.2\mathrm{OP} + 1.2\sum \mathrm{SL} + 1.0W$

$0.9D + 0.9A + 1.2\mathrm{OP} + 1.2\sum \mathrm{SL} + E$

$0.9D + \mathrm{TP} + 1.2\sum \mathrm{SL}$

5.7.3 Long-Duration Load Combination at Temperatures within the Creep Range

This LRFD load combination includes sustained loads only and is considered a long-duration condition. The resulting required strength should be compared to a design strength determined with a limiting creep stress based on 100,000 h load duration.

$1.4D + 1.4A + 1.4OP + 1.4\sum SL$

5.7.4 Short-Duration Load Combinations at Temperatures within the Creep Range

The following LRFD load combinations include sustained plus transient loads and are considered short-duration conditions. The resulting required strengths should be compared to a design strength determined with a limiting creep stress based on an appropriately short cumulative duration.

$1.2D + 1.2A + 1.2\mathrm{OP} + 1.2\sum \mathrm{SL} + 1.6L + 0.5S$

$1.2D + 1.2A + 1.2\mathrm{OP} + 1.2\sum \mathrm{SL} + 1.6S + (L \text{ or } 0.5W)$

$1.2D + 1.2A + 1.2\mathrm{OP} + 1.2\sum \mathrm{SL} + 1.0W + L + 0.5S$

$1.2D + 1.2A + 1.2\mathrm{OP} + 1.2\sum \mathrm{SL} + E + L + 0.5S$

$1.2D + 1.2A + \mathrm{TP} + 1.2\sum \mathrm{SL} + 0.5L + 0.2S$

$0.9D + 1.2\mathrm{OP} + 1.2\sum \mathrm{SL} + 1.0W$

$0.9D + 0.9A + 1.2\mathrm{OP} + 1.2\sum \mathrm{SL} + E$

$0.9D + \mathrm{TP} + 1.2\sum \mathrm{SL}$

5.7.5 Load Combination at Excursion Temperature

This LRFD load combination includes sustained loads at the excursion temperature. Design strengths, reduced for the temperature effects, shall apply.

$1.2D + 1.2A + 1.2\mathrm{OP} + 1.2\sum \mathrm{SL}$

5.7.6 Thermal Gradient Load Combinations at Operating Temperature

The following LRFD load combinations include thermal loading conditions at the operating temperature. These load combinations are usually considered only if the structural engineer expects significant thermal loads. Design strengths, reduced for the temperature effects, shall apply. If the operating temperature is within the creep range, a design strength determined with a limiting creep stress based on 100,000 h load duration shall apply to the first load combination, and a design strength determined with a limiting creep stress based on an appropriately short duration shall apply to the second load combination.

$$1.2D + 1.2A + 1.2\mathrm{OP} + 1.2\sum \mathrm{SL} + 1.2\mathrm{TL}$$

$$1.2D + 1.2A + 1.2\mathrm{OP} + 1.2\sum \mathrm{SL} + \mathrm{TL} + L + 0.5S$$

References

AISC. 2017a. *Specification for structural steel buildings.* AISC 360-16. Chicago: AISC.

AISC. 2017b. *Steel construction manual.* AISC 325-16. Chicago: AISC.

ASCE. 2017. *Minimum design loads and associated criteria for building and other structures.* ASCE 7-16. Reston, VA: ASCE.

CHAPTER 6

Ductwork Global Structural Analysis

6.1 INTRODUCTION

Air and gas ducts and their support structures are major structures that require both global and local structural analysis to predict their performance and verify their structural integrity. Local structural element design and stress evaluation are discussed in Chapters 7 and 8. This chapter discusses factors to be considered and methodologies to be used in the global analysis of ductwork and its support structure.

Many factors influence the structural design and global evaluation of a ductwork system. The structural engineer must have insight into the expected behavior of the duct and its support structure under the various operating, excursion, and transient conditions. Construction issues such as erection sequence, crane capacities, temporary support requirements, and fabrication and erection tolerances should be considered. The behavior of the soil and the foundation should be accounted for, along with any interaction with equipment and other structures. The structural engineer must be aware of how all these factors combine to influence the behavior of the duct and its support system.

The ductwork layout and support scheme should result in a system with predictable load transfer paths. This is discussed in detail in Chapter 2. Once this scheme has been established, the structural analysis task is typically straightforward. A rational analytical approach is required, using a proven methodology, realistic boundary conditions, and reasonably accurate section properties. What is required is an analysis that produces internal forces and support reactions that can be used to complete the structural design of the individual ductwork elements. In addition, displacements from thermal expansion and structural deflection are usually needed at interface points, such as expansion joints and dampers, to confirm the functionality of the ductwork and its support system.

6.2 GLOBAL APPROACH

The structural analytical approach adopted may consist of classical methods (manual calculations), computer analysis, or some combination thereof. The method chosen should be dictated by the complexity of the duct or support system and the degree of accuracy appropriate for the type of structure being designed.

Sizing and routing of ductwork are typically performed by the process mechanical engineer. However, the structural engineer should participate in the early stages of ductwork layout, especially the duct support arrangement. The structural engineer's contributions would include advice on expansion joint and support spacing and location, early assessment of material quantities, and the structural consequences of system-driven and interdisciplinary decisions. There may also be alternative arrangements that may have structural benefits not readily identified by other engineers. See Chapter 2 for more on duct system sizing and arrangement.

6.2.1 Decoupling of Supports and Ductwork

The general practice in the ductwork structural engineering industry is to analyze and design the ductwork and its support structures separately. In this process, the configuration of the ductwork system is developed as described in Chapter 2 with emphasis on simplifying the support system while meeting the system functional requirements. The intent should be to create an arrangement that allows a determinate load path and permits decoupling from the ductwork and support system designs. Decoupling the ductwork from the supporting system could avoid unexpected stresses for a wide range of temperature change.

This is a relatively straightforward process when the ductwork is routed inside another structure, such as a boiler building. In this situation, the structure provides support to virtually any load applied by the ductwork without significant interaction, such as deflections or settlement, that might result in load redistribution. External ductwork can usually be treated similarly; however, special considerations for interaction are discussed subsequently in this chapter.

6.2.2 Load Paths

Fundamental in the analytical process is the need to verify that paths exist to transmit all external and system loads down to the foundation. Careful assessment of the structure's behavior is needed to confirm that the structure will function as designed. This is true for individual structural elements, as well as the connections among the major structural subsystems.

In the decoupled approach, the structural engineer designing the ductwork must establish various load paths within the ductwork structure that will carry all applied loads to the duct support structure. The load paths must be chosen or determined using the various ductwork structural elements described and discussed in Chapters 2 and 6. These structural elements consist of internal trusses that stiffen the global structure and bridge between the duct supports, plate panels, corner angle stiffeners, and duct plate stiffeners or other elements needed for this purpose.

In a similar fashion, proper load paths must be chosen or determined in the duct supporting structures so that the loads and forces make their way into the earth.

Ductwork. Preliminary sizing of the ductwork structural elements should be based on knowledge of the operating and transient loads and the ductwork system operating conditions. These items are discussed in Chapter 4. Chapters 7 and 8 present discussions of typical ductwork stiffener layout and sizing considerations. Although final details of the ductwork configuration are not needed at this time, some consideration of these details is needed early to ensure that the analytical load paths are representative of those in the actual structure.

As indicated in Figures 6-1 and 6-2, these structural elements must be arranged to provide a load path to the support structure. Figure 6-1 shows a loading concept for resisting dead, live, and gravity loads in rectangular ductwork. In this situation, the uniformly distributed vertical loads on the top and bottom plates must be transmitted to the side walls and then out to the supports. To accomplish this load transfer, stiffening of the top plate is necessary to transmit the loads to the walls. Then the side panels carry the loads to the supports. In a similar fashion, other vertical loads, such as ash accumulation on the bottom plate or in a hopper or live load on the duct top, are carried to the duct supports. Where a simple span is used, the reactions are readily determined and assumed to be applied to the support structure without interaction. However, for long duct runs, the situation could benefit from a continuous duct spanning several supports. In this case, an indeterminate classical solution would be required to determine the

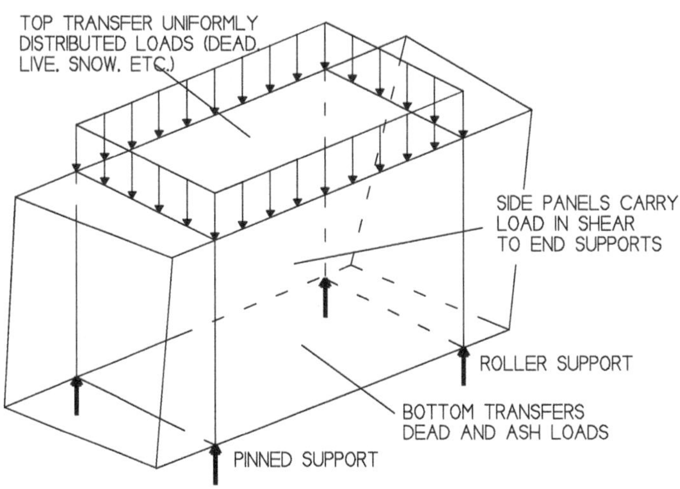

Figure 6-1. Typical load path for vertical loads.

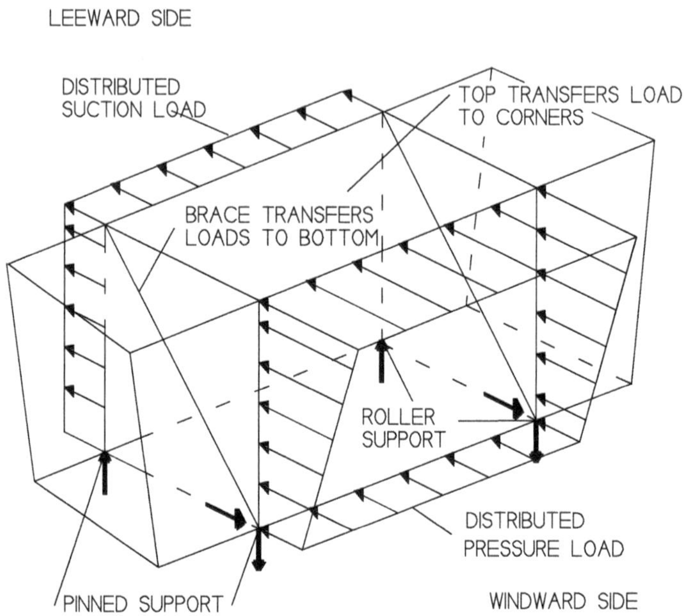

Figure 6-2. Typical load path for wind loads.

reactions. In an indeterminate solution, the stiffness of the supporting structure enters into the distribution of the duct loads to its various supports.

Figure 6-2 shows the typical load path for wind loads. In this case, the distributed wind loads on the stiffened sides must be carried to the top and bottom corners. From here, the top and bottom plates transfer the loads to the supports in diaphragm or truss action. An internal truss or rigid frame is also required to transfer the upper plate wind loads through the rectangular ductwork to the support structure while maintaining the ductwork configuration. In a similar fashion, other horizontal loads, such as horizontal seismic or unbalanced pressure forces, are carried to the duct supports.

System loads must also be resisted. The major system loads are the static pressure, unbalanced pressure, ash, and major equipment supported by the ductwork. See Section 4.2 for discussion of all design loads. For each type of load, a load path must be established and used to transmit the loads from their origins to the support structure. For example, consider a ductwork section that cantilevers over a support and turns 90°, ending at an expansion joint, as shown in Figures 4-1 and 7-6. In this situation, the unbalanced pressure produces loads on the support and bending in the ductwork that can result in more than one support being involved in resisting the loads.

Load path identification is more straightforward in circular ductwork. Provided that the ductwork material has been properly sized (as discussed in Chapter 7), the circular shape facilitates the transfer of applied loads to the support structure in an efficient manner as a single structural element. The details of the

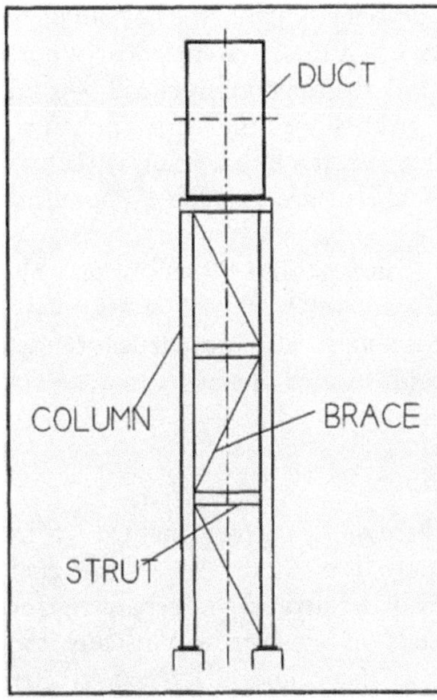

Figure 6-3. Typical external support structure.

duct support attachments to the supporting structure, which ensure the continuity of the load paths, are equally important. Internal trusses or rigid frames at supports could maintain the ductwork configuration and stability, which relates to the critical buckling stress discussed in Section 6.3.3.

Supporting Structures. With the ductwork and its support structures decoupled, the support analysis and design is straightforward. With the framing geometry previously established, preliminary member sizing can be performed using simple approximations. A reasonable approach for preliminary member sizing for the support structures is to (1) estimate the dead, live, and wind loads; (2) calculate preliminary column loads; (3) perform a decoupled analysis as described; and (4) estimate the member sizes using the current version of *Specification for Structural Steel Buildings,* ANSI/AISC 360 (AISC 2017). An example of a simple duct support structure is shown in Figure 6-3. Braces and struts are normally considered to have pinned connections, so columns, braces, and struts can be easily sized as axially loaded members. If the ductwork is within a large structure, such as a boiler room, sufficient load paths usually exist. However, the addition of local bracing might be necessary. The support structure's horizontal and vertical bracing configuration shall be such that all lateral and longitudinal loads from each duct support are transferred to the foundation.

6.2.3 Load Interactions

From the preceding discussion, it should be obvious that structural elements in ductwork perform more than one function simultaneously. For example, stiffeners used to transfer ash and dead loads across the bottom plate to the side wall trusses might also be used as part of a truss system to carry horizontal wind or seismic loads to the supports. Similarly, the bottom chord of a side wall Vierendeel truss will also participate in horizontal load transfers. The structural engineer should be familiar with the multiple functions that many structural elements perform in ductwork design and should ensure that the proper strength is provided in the design.

For determinate structures, stiffness does not play an important role in load path determination. The loads will be carried as assumed, because any redundant capacity will be a secondary effect. This is not true for indeterminate structures. In that situation, a more sophisticated approach may be appropriate in evaluating ductwork load paths.

6.2.4 Classical Methods

Classical methods as discussed here are manual or hand calculations using traditional structural analysis methods. When the ductwork and its support structures are decoupled or treated as two separate structures in the structural analysis, classical methods of analysis will usually be sufficient. A classical approach can be used to determine all structural element forces when the structures are determinate. Trusses can be solved using free body diagrams, and the member loads will be axial, with little or no moment interaction. Wall, floor, and roof panels can be treated as trusses or girder webs using classical methods for calculating shears and moments.

Structures or portions of structures that are indeterminate in a single plane can also be analyzed using classical methods such as moment distribution and virtual work. These methods will produce moments, shears, and reactions that should be considered in the design.

In the preliminary stages of design, classical methods should also be used to give the structural engineer a better understanding of the important characteristics of a ductwork system that affect the overall structural behavior, even if computer techniques will ultimately be implemented in the final detailed analysis and design. This provides confidence in the design that may prove important as the final detailed design evolves.

6.2.5 Computer Models

Physical plant constraints and system functional requirements can result in ductwork configurations that may require a more sophisticated analytical approach. This is particularly true where the assumption that the ductwork and its supports can be analyzed separately is questionable. In these cases, computer modeling and analysis may be beneficial and prudent. Following are examples of cases where computer analysis may be performed:

- When the configuration of the system is such that stability is a major concern or when the ductwork or the support system is statically indeterminate;

- When there are intermediate phases in the construction sequence where an alternative load path is needed or when exposure to severe environmental loads on the partially completed structure is significant;
- When unusual foundation-related restrictions have been identified that might warrant special attention during the analysis—such a situation can arise when uplift loads must be limited because of difficult soil conditions that limit pile capacity or when significant differential settlement is expected and unavoidable;
- When physical constraints imposed by the need for strength and stability result in undesirable behavior because of operating or excursion differential temperatures within the duct;
- When local codes or customer specifications require response-spectrum analysis for a seismic event; and
- When a complex, redundantly supported ductwork design is being used or when the relative stiffness of structural elements results in load redistribution under some load combinations, such as when the lateral stiffness of the ductwork is adequate to transfer loads to interfacing equipment rather than allowing it to be resisted by the supporting structure.

The presence of these factors could indicate that computer modeling and analysis of the ductwork and its support system is the most efficient method to develop the information needed for the design. Further, although classical methods provide the structural engineer with enough information for member sizing, the need for selected deflection data to ensure functionality at critical interfaces, along with other issues discussed earlier, may compel the structural engineer to use computer solutions. Used within their known limitations, computer analysis and design can provide a more efficient evaluation and design.

6.2.6 Finite Element Analysis

Finite element analysis is a numerical method to find approximate solutions for partial differential equations. It divides the ductwork structural system into smaller finite elements such as beam, plate, shell, and solid. For simpler finite elements, the equations can be described easily and then be assembled into a combined matrix of equations. Equations for beam elements are more complex than those for shell or solid elements, so fewer elements are required for acceptable results. For example, the web plate of rectangular ductwork should include many shell elements to model shear variation from top to bottom, because most shell element equations are based on uniform shear distribution; however, a beam element for stiffeners could describe linear variation of shear because of element equations. An acceptable computer model for flexural analysis requires more shell elements to capture shear lag effects between the top or bottom plate and the web plate. Height ratio or height-to-thickness ratio, in addition to the flange width to span length ratio, has a significant influence on shear lag.

Ductwork is a geometrical nonlinear system, because the thickness of duct plate is thin compared with the span. The duct plates require nonlinear analyses for tension field action (Section 6.3.2) for rectangular ductwork, and a higher safety factor for circular ductwork (Section 6.3.3). For geometrical nonlinear systems, deformations of structures are to be used to modify the stiffness of the structures in analysis through an iteration procedure. Although the higher buckling safety factor for circular ductwork is a geometric nonlinear phenomenon, it could be derived from eigenvalue-based buckling analysis based on a linear finite element model system.

Ductwork could be simplified as finite element analysis of a three-dimensional (3D) truss or fictitious model, as stated in this chapter, to avoid shear lag, tension field action, or other geometry-related issues, as well as to obtain a quicker solution.

6.3 STRUCTURAL MODEL CONSIDERATIONS

Independent of the analytical approach, the structural engineer must conceptualize and create a structural model, or free body diagram, of the ductwork and support structure that is to be analyzed. This model should be an accurate approximation of the actual structure and represent the structure's functional capabilities. The methods used to develop critical elements of the model and the application of the loads are discussed in the following subsections.

6.3.1 Support Structures

Modeling the supporting structures for a ductwork system is straightforward. A realistic two-dimensional (2D) or 3D model, such as that shown in Figure 6-4, should be developed. The truss elements, braces, and struts should have three degrees of freedom per node. Columns should be represented by continuous beam elements to account for the continuity bending moments and for a proper stability assessment. The details of geometry, member orientation, member properties, and work points require close attention.

Computer models should not include connection details; connections are typically represented in the model by member end releases. Connections are assumed to be fixed, to be free to rotate, or to have a rotation spring stiffness that must be included in the model. Connection angles, flange restraints, and member stiffeners need not be modeled.

In creation of the model, consideration should also be given to how the loads will be applied. For example, early decisions should be made regarding which loads should be applied as uniform loads on surfaces, line loads on members, or point loads on members or at joints.

6.3.2 Rectangular Ductwork

Special care must be taken when creating the ductwork structural model. Provision must be made to position loads correctly so that support forces are

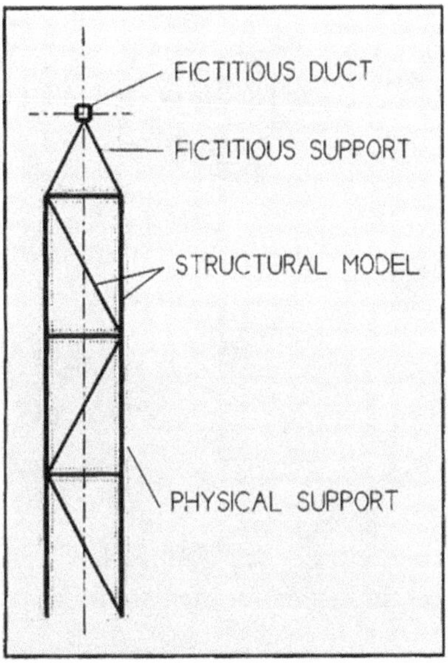

Figure 6-4. Typical support model.

accurately determined. Care also must be taken to assure that proper continuity is maintained in the ductwork structure model. The modeling techniques described here may allow unrealistic flexibility in the longitudinal direction, which can result in faulty load distribution in the analysis. Furthermore, the presence of internal bracing, longitudinal stiffeners, or fixed-end transverse stiffeners may substantially increase the ductwork's assumed sectional properties.

As with the support structure model, the ductwork analysis model should not specifically include details such as local stiffeners, connections, or internal flow distribution devices. The design of these items is covered in Chapters 8 and 9. Their inclusion typically will not affect the results of the global analysis. Furthermore, the design of these items is more easily completed after the analytical results are available.

Axial and Flexural Considerations. In rectangular ducts, because of the plate's high width-to-thickness or b/t ratio and high height-to-thickness or h/t ratio, the duct is usually not as stiff as a typical box beam. Thus, only a portion of the ductwork should be considered when calculating the ductwork stiffness in resisting bending loads from shear lag effects. For this purpose, the effective section is initially based on a plate width of $16t$ from the edge of any corner stiffener (Figure 6-5). See Chapter 8 for more on the effective width of duct plate and the composite action between stiffener and plate. Any longitudinal stiffeners and corner angles that are provided to reinforce the ductwork corners should also be included in the stiffness calculation. This effective flange and web area is then used to derive axial and flexural member properties. The torsional property of ductwork is based on the cross section of the box and the thickness of the wall,

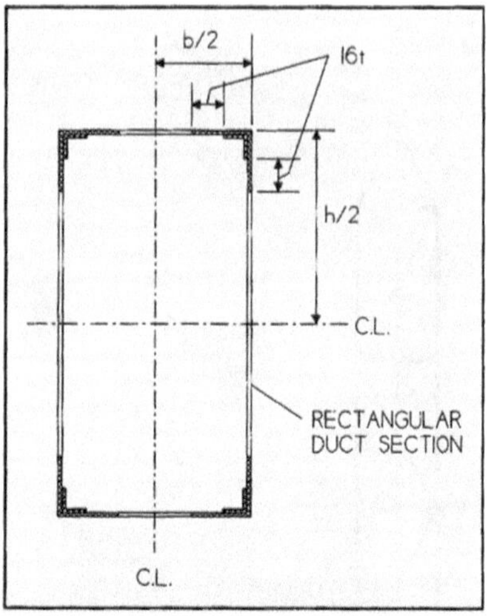

Figure 6-5. Effective area considered for duct stiffness.

which may not be reduced by shear lag effect, but by tension field action, discussed in the next subsections.

The need for turning vanes or other flow distribution devices will complicate the ductwork's structural behavior. Changes in direction, where trusses and turning vanes are usually located, are often associated with a congested stiffener pattern. Therefore, these areas are quite rigid. In these locations, it is often appropriate to use the full section properties of the ductwork so that indeterminate conditions are properly considered. Where the course of action is unclear, it is prudent to analyze a range of properties to envelop all possible behavior.

When stresses in the duct wall theoretical webs and flanges are low, the corner elements become more effective. The assumed initial structural element values, primarily thicknesses, can be modified based on results in accordance with the design, as discussed in Chapters 3, 7, and 8.

Shear Considerations. For shear, the side walls and the top and bottom panels of a duct can be modeled in two different ways. First, they can be analyzed using truss methodology, with the stiffeners acting as struts, the duct plate as the diagonal members, and the corner angles as the top and bottom chords of the truss. The second method assumes that the duct plate is the web of a large fabricated girder, with the panel stiffeners as the girder web stiffeners. The effective corner areas discussed in Section 6.3.2 become the girder top and bottom flanges.

The decision to model a plate wall as a girder or a truss should be made based on the b/t ratio or h/t ratio of the plate as applicable. Girder web design is based on a combination of both web shear and tension field action between the web stiffeners (Figure 6-6). Both mechanisms are assumed to contribute to the girder strength. Shear load is initially resisted directly by the plate. However, as the load increases, the plate is assumed to buckle between the stiffeners along a 45 degree

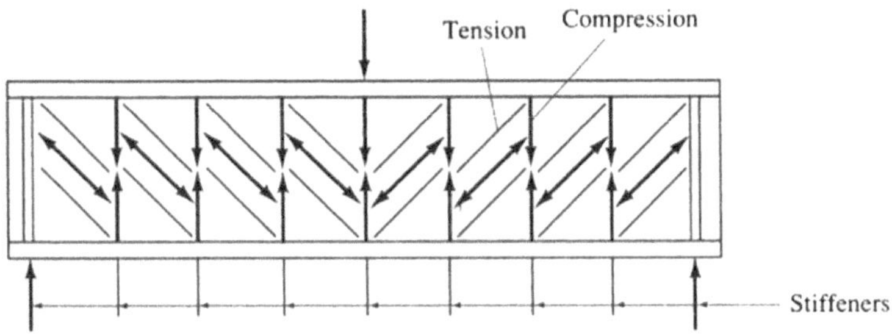

Figure 6-6. Tension field action.

direction, with the remaining load being resisted by plate at 90 degrees to the plate buckling direction through tension field action. Therefore, tension field action provides smaller capacity compared with web shear. Web plate buckling depends on the material's modulus of elasticity, E, among other things. With increasing temperature, E decreases, as does the plate's buckling capacity. The plate's pure shear capacity is also decreased, but not directly proportionate to the buckling capacity. Thus, the reduction in total shear capacity with increasing temperature is not linear. The actual reduction in strength for a particular configuration depends on the aspect ratio of the stiffened panel and the slenderness of the plate. The structural engineer should refer to the theory on this subject in a textbook, such as Salmon et al. (2008), to develop plate shear capacity tables similar to those in the AISC specification.

A finite element model can use the actual plate elements instead of actualizing the plates into elements. Shear lag effect and tension field action can only be captured through analysis using a very small mesh of shell elements. Tension field action can only be captured through a nonlinear iteration procedure. For circular ductwork, the design basis of cylinder buckling is also a result of nonlinear analysis or eigenvalue analysis.

For long duct spans where significant shear could accumulate in the plate near the supports, it is usually more economical to add diagonal stiffeners to the plate to carry the shear force rather than increasing the plate thickness.

6.3.3 Circular Ductwork

The cross section of circular ductwork is normally considered fully effective. However, depending on the final configuration and the methods used to preclude cylinder buckling, the strength may be quite low. This is a result of imposing high safety factors against the critical buckling stress. Therefore, the details improving buckling stress, such as a diaphragm or a truss at supports, may govern circular ductwork design.

For modeling purposes, it is reasonable to use the full duct cross section when determining a model's stiffness. On the assumption that the support details can be designed to accommodate the loads without local buckling, the load distributions should be somewhat insensitive to subsequent changes in ductwork stiffening.

6.3.4 Modeling Strategy

When a coupled analysis of the ductwork and its support structure is desired or required by code, such as the combined analysis required by ASCE 7 (2017, Chapter 15) for seismic load analysis of nonbuilding structures supported by other structures with vertical mass irregularities, the structural engineer is to create a model configuration that is representative of the physical ductwork with its applied loads. The most obvious approach is to model the ductwork geometric shapes with plate or shell elements. This provides the model surfaces over which the loads can be applied, but there are some difficulties.

- The stiffness of a model constructed of plate elements will not be similar to the stiffness calculated using effective area, because of shear lag analysis and tension field action. Unless the material properties are adjusted, the model will be stiffer than the real structure.
- Internal load transfers, primarily from the surfaces to the supports, will be unpredictable in the model unless all the internal ductwork elements are modeled.
- Connection behavior between the ductwork and its support structure will be complicated in the model, because many supports have gaps or slide bearing plates, which may cover multiple elements. It is critical that the member releases or specifications are consistent with the support details to ensure that the load transfer between the duct and the support structure is captured correctly.
- The analysis model can become too large and cumbersome to be used effectively.

An alternative modeling approach is to develop a fictitious ductwork element positioned at the centroid of ductwork resistance (Figure 6-7). This ductwork element can be positioned by using fictitious rigid links to connect it to the support

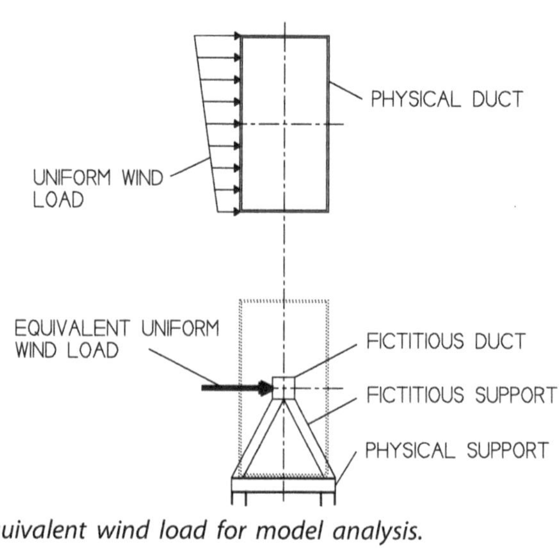

Figure 6-7. Equivalent wind load for model analysis.

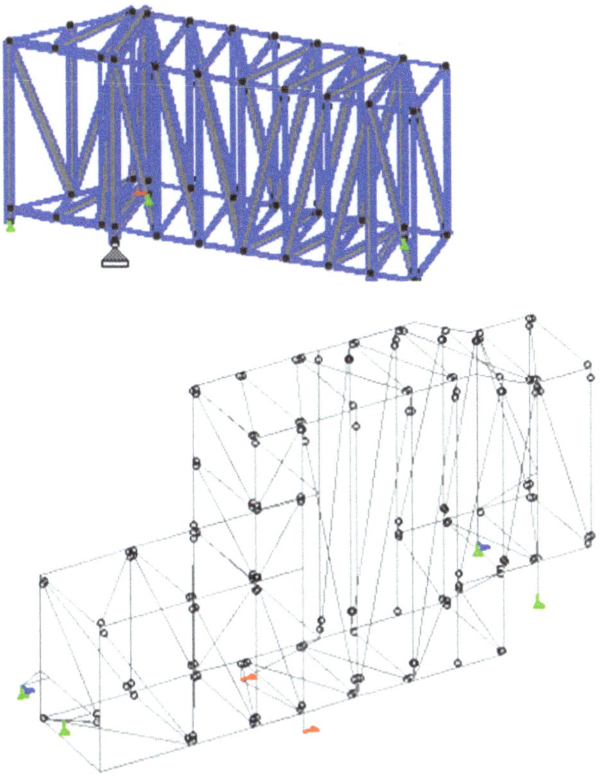

Figure 6-8. Equivalent 3D truss model analysis.

structure in the model in the same manner as it is really connected. Uniform loads can be consolidated into point or line loads for application to the simplified model. This approach also reduces the model size, which reduces post-analysis work and improves confidence in the results.

Another alternative approach is to develop a fictitious 3D truss model or a series of 2D models (Figure 6-8). The corner members include the area of corner angle plus effective plate sections per shear lag effects. The vertical web members include vertical stiffeners plus effective section per shear lag effect. The inclined members are tension-only elements per size of plate. This model can provide more realistic results for complicated ductwork configurations. This approach may be simplified by using an adjusted stiffener spacing. Not all vertical and horizontal stiffeners need to be modeled. The forces can be extracted from the 3D model and used to design the duct stiffeners, duct plate, corner angle, and supports so that all local effects may be correctly combined with the global effects.

6.3.5 Interface Boundary Conditions

There are several interface points between the ductwork, support structure, and adjacent structures where boundary conditions must be assessed for proper inclusion in the structural analysis model. Details of the connections between the components of the supporting structures, and with the ductwork itself, must be conceptualized to determine how each type of load is transferred. The goal is to faithfully represent structural functions that affect the interaction between the ductwork and its support structures so that the response is correctly predicted.

Foundations. Foundations are normally considered as rigid; they are usually assumed not to rotate or displace under applied loads. This is usually reasonable unless site soil conditions dictate otherwise.

Differential foundation settlement of determinate structures does not normally present a problem. The structure will accommodate significant settlement with minor joint rotation. An indeterminate structure should be evaluated more closely. Ductwork that crosses several supports continuously should be considered indeterminate. For such structures, uneven foundation settlement will induce bending in the ductwork that could redistribute the support loads and possibly affect the ductwork design itself.

Other interactions are possible in poor soil. For example, piles in poor soil may have adequate vertical capacity but develop significant horizontal displacements, and thus inadequate horizontal capacity, under lateral loads. These situations should be evaluated to determine their effects on the ductwork system and its interfaces with equipment.

Interface between the Foundation and the Duct Support Structure. The connection to the foundation for columns is normally considered to be pinned. Rigid connections may be of some value when the displacements of moment-resisting frames are a problem or where the equipment arrangement prohibits using braced frames.

Interface between the Support Structure and the Ductwork. There are many types of duct support components, as discussed in Chapter 2. Slide bearing plates under support posts are normally considered as theoretical 3D rollers; they are only vertical supports and are free to move in all horizontal directions. Slide bearing plates with guide bars are normally considered as rollers in the direction parallel to the guide bars and as pins in the direction perpendicular to the guide bars. Rather than attempting to perform an exact friction analysis, it is reasonable to apply estimated friction loads opposite to the direction of thermal expansion at the location of the slide bearing plates.

Hanger rods are treated as theoretical rollers in a free body diagram. Some types of hangers require special attention. For example, a constant-support hanger should be modeled as a force rather than a support, and a variable-support hanger should be modeled as a spring. A rigid rod always acts as a hard support. With hangers, mathematical instabilities will occur in the model unless lateral load transfer mechanisms in the form of guides, bumpers, or stop bars are included.

Interface between Duct Segments. Nonmetallic expansion joints, which are used to connect sections of ductwork to equipment or other sections of ductwork are normally considered a free boundary condition. This is reasonable because of the negligible stiffness and strength of these joints.

Metal expansion joints are usually modeled as springs. They have appreciable axial and rotational spring stiffness. Depending on how they are installed, they could even have preload forces. The spring coefficients can be obtained from the joint manufacturer. The joint detail should be evaluated to determine whether lock-up is possible, especially in reassessment work. In that case, the analysis should also consider the possibility of fixity at that location.

Shear keys, which are often present in toggle ductwork sections, are treated as translational restraints. Care must be taken when modeling such components, given the wide range of geometries. It is best to avoid the use of gap finite elements in this type of analysis. Nonlinear analysis is unjustified in this situation. Provided that the joint functions properly, gap interactions should not affect the predicted load distribution.

The restraints at each interface can be included in the model as described. A complete tabulation of the connection and support releases facilitates data input and later validation. It also provides an easy method of demonstrating that individual and combined loads can be accommodated.

Interface with Hoppers. The additional mass that results from the storage capacity of ash collection hoppers should be included in the structural analysis model. Distribution of this potentially large load can be affected by model stiffness.

In some situations, ash hoppers can impose local thermal loads on the ductwork that should be evaluated to determine whether they may be significant. These arise because of both the insulating characteristics of ash during operation and its heat storage capability during shutdown.

6.3.6 Load Path Assessment

The assessment of load path performance is straightforward when the analysis is performed manually for decoupled systems, as described in Section 6.2.2. Decisions are made regarding the load path, and analysis proceeds on that basis. Faults seldom occur, because the assumed load path is created as part of the preliminary design. Conversely, when computer modeling is used, the load path in an analytical model is sometimes difficult to verify, which can introduce inconsistencies. This emphasizes the need for an inherent understanding by the structural engineer of the basic structural system and all subsystems. Before computer modeling, preliminary hand calculations should be performed to determine initial sizes and stiffeners and to develop an understanding of the system behavior.

6.3.7 Load Development and Application

When classical methods are used for structural analysis, the development of loads requires no special provisions. The loads are applied directly to the structural elements as the various parts are evaluated.

When performing a computer structural analysis using an analytical model that approximates the structure, as described in this chapter, attention must be given to how the loads are applied. Specifically, the following loads may require special treatment.

Ductwork Dead Load. When the ductwork is structurally modeled as described in this section, as a fictitious member with calculated properties positioned at the centroid of the ductwork shape, rather than a prismatic section, automatic self-weight determination by a computer program will not be correct. Modification of the material density or application of the duct load as a uniformly distributed load are two methods to include the total weight of the actual duct and all its attachments. Flow distribution devices and other ductwork accessories and attachments can be best treated as an applied concentrated load.

Ash Load. In a detailed model, ash loads can be applied to the bottom ductwork plate, as is expected in the real structure. Load transfer would be accomplished by the plate elements. Using the fictitious element approach to position the ductwork element will cause all ash loads to be applied at the point where the rigid links connect to the support structure. In both cases this is not exactly representative of the real structure. An assessment must be made to determine how the ash load will be transferred to and carried by the support structure. Details regarding the load distribution at the local level are provided in Chapters 4 and 8.

It is common practice to suspend hoppers from ductwork, so it is also necessary to include the ash load from a full hopper in the loadings. On the other hand, if a separate support arrangement is provided for hoppers, unbalanced pressure load effects should be investigated.

Wind Load. Wind loads are normally applied over the entire ductwork projected area. However, this may be simplified for the support structure by applying wind as joint loads. The wind load derivation should be based on the projected area of the ductwork with insulation and lagging. For larger sizes of rectangular ductwork, wind loads may follow the building code for applied windward, leeward, upward, and downward directions.

Thermal Differentials. Thermal differential loads typically only apply to the ductwork. The support structure is not normally subjected to this loading unless the support conditions are such that equal and opposite reactions are not developed. The normal strategy is to provide support and interface conditions that allow thermal movements to take place unimpeded. Static friction forces toward the fixed point of the ductwork module should be considered in 3D ductwork analysis.

Thermal gradients can arise from the insulating properties of ash as it collects on the bottom of ductwork and in hoppers. There have been situations where thermal gradients have imposed significant loads, and corresponding deflections, in structures. The structural engineer should be alert to this possibility as the model is developed and evaluated.

One way to apply thermal loads to an integrated computer analytical model is to permit the support structure material to have a coefficient of thermal expansion specified as zero while the ductwork has the correct value for the materials being used. Thermal gradients within a duct structure are handled by assigning the expected absolute temperature to each individual element, including all the plate elements.

Seismic Loads. For outside ductwork, the environmental load combinations with wind loading typically controls over the load combinations with seismic loading. However, in areas of high seismic demand or where the ash loading is large, seismic inertial forces can have important consequences for structural integrity. In seismic design categories D, E, and F, response-spectrum analysis may be specified for code compliance or chosen by the structural engineer to reduce the forces that result from simplified approaches, such as pseudostatic analysis.

Per ASCE 7, Chapter 15, ductwork is a nonbuilding structure supported by another structure, and three approaches may be used to analyze seismic forces based on properties of the ductwork and supporting system.

- *Approach 1 –Ductwork weight less than 25% of the combined weight of ductwork and supporting system.* The seismic forces should be determined as components in accordance with ASCE 7, Chapter 13. The support structure seismic system should follow either Chapter 12 or Chapter 15.
- *Approach 2 – Ductwork weight 25% or more of the combined weight of ductwork and supporting system, plus a fundamental period for the ductwork of less than 0.06 s.* The ductwork is permitted to be designed as components in accordance with ASCE 7, Chapter 13. The support structure has a seismic system following either Chapters 12 or 15.
- *Approach 3 – Ductwork weight 25% or more of the combined weight of ductwork and supporting system, plus a fundamental period for the ductwork of 0.06 s or longer.* The ductwork and supporting structure shall be modeled together, in a combined model, with appropriate stiffness and effective seismic weight distributions. This can be done similarly to the combined ductwork and support structure model in Figure 6-7 or by using spring supports representing the support structure stiffness.

Unbalanced Pressure. Pressure loads on closed sections of ductwork are internal forces that cancel out, and so need not be included in the overall structural analysis. However, opposite expansion joints in open sections of ductwork, or at connections to equipment, the internal pressure produces an external force between duct sections that can be significant and should be considered in the ductwork and support structure design.

6.3.8 Model Validation

Before proceeding with the final design of the individual structural elements, it is prudent to assess the accuracy of the analytical results. The results should be validated for correctness. Conformity with the expected performance should also be assessed. For a computer model, this begins with the obvious first task of verifying the accuracy of the model input, boundary conditions, and loading information. It is prudent to compare the support reactions in the computer software analytical results to reactions computed using other means as a way to check and assist in troubleshooting erroneous computer model analytical results.

For indeterminate structures that may have external constraints, it is useful to begin the analysis with a simplified structure. These results then can be used to baseline the final configuration analysis. This approach provides a good approximate check of the final results and, if the constraints are applied sequentially rather than all at once, offers a method for debugging the computer model.

6.3.9 Member Forces and Reactions

Member internal forces and moments are needed to perform the final member sizing and to design connections. These should be calculated or obtained at intervals along the members that will ensure that the maximum component combinations have been enveloped. The accuracy of automated member sizing depends on this.

Support reactions should be evaluated from several viewpoints. First, the sum of the reactions must approximately equal the sum of the known loads for each loading combination. This simple check is fundamental to assure model accuracy and to ensure that the loads have been properly considered. Although input loads and reactions can be compared on request by some computer software, this is not an independent check, and it cannot ensure that fictitious or erroneous loads have not been introduced. This automated method also will not determine whether loads were incorrectly entered in the model when it was created. Second, the resulting reactions should be evaluated to determine that they are within the capacity of the foundations. Mismatches at this stage in the analysis and design effort should be corrected before proceeding with final member sizing. Finally, overturning and sliding of the structure should be evaluated. It is prudent, at this point, to check the results of all loading combinations for situations that result in net uplift at supports.

Net uplift is often the result of cantilevered-type ductwork configurations or overturning moments from lateral loads such as unbalanced pressure, seismic, wind, and friction. A cantilevered arrangement that results in permanent dead-load-driven uplift or lateral forces should be avoided where practical. It is always good practice to modify the support scheme to eliminate net uplift, if possible, before proceeding further with the design.

6.3.10 Displacements

Large unexpected displacements in the computer analysis output are clear indications that the model is not correct. These should be resolved before proceeding. Most often they are the result of an extraneous member release, excessive load, or incorrect member property.

However, a more disciplined displacement evaluation of the computer output should be undertaken in indeterminate structures with potentially redundant load paths to identify more subtle potential model errors. For example, an inadvertent release at one support leg could require the ductwork model to redistribute loads to other supports. This could cause locally undersized members and erroneous loads to be applied to the support structure. In addition, ductwork forces elsewhere could be significantly increased.

With thermal loads included in the analysis, thermal movements should be reviewed at ductwork interfaces to ensure that the behavior of connecting accessories, such as expansion joints, is not compromised. This would include an assessment of the thermal movement at the slide bearing supports to assure proper functioning. The thermal movements will also be important final details in the design of the duct support structure elements for eccentric support loadings.

References

AISC. 2017. *Specification for structural steel buildings.* AISC 360-16. Chicago: AISC.

ASCE. 2017. *Minimum design loads and associated criteria for building and other structures.* ASCE 7-16. Reston, VA: ASCE.

Salmon, C. G., J. E. Johnson, and F. A. Malhas. 2008. *Steel structures: Design and behavior.* New York: Pearson.

CHAPTER 7
Plate Design

7.1 INTRODUCTION

There are many possible duct plate configurations. Duct plate can have different support conditions, stiffener positioning, and edge distance constraints that affect the plate design. Both good references and sound engineering judgement should be used by the ductwork structural engineer as each design decision presents itself. Overall project cost, performance, and constructability should always be considered when making decisions regarding the basic duct shape, plate thickness, and stiffener arrangement.

Chapters 4 and 5 present the types of loads, load combinations, and associated design strengths that are to be used for ductwork panel design. These loads and load combinations are used to calculate the forces within the duct plate, the deflection, and the dynamic response.

The material in this chapter should be used concurrently with the information in Chapter 8 because of the interaction and composite behavior between the duct plate and the stiffeners.

7.1.1 Global Behavior

The duct plate is a crucial component of the overall structural system, because it ties all the other components (duct stiffeners, corner angle, supports, etc.) together.

The in-plane stiffness of the roof and floor duct plate is large relative to the support frames, and therefore the in-plane deformation of the duct plate does not have a significant impact on the overall story drift (Figure 7-1). Regardless, the duct plate is typically idealized as a flexible diaphragm similar to untopped steel decking per Section 12.3.1.1 of ASCE 7 (2017). Lateral loads are typically distributed to the support frames, independent of the support frame lateral stiffness, consistent with a continuous beam analogy. A more detailed analysis that addresses the relative lateral stiffness of the support frame and the duct plate may be performed at the discretion of the structural engineer if there is concern that a flexible diaphragm assumption may be unconservative or if a more accurate analysis is required.

There is typically not a significant difference between the vertical stiffness of the wall duct plate and the duct support frames. The vertical loads are distributed

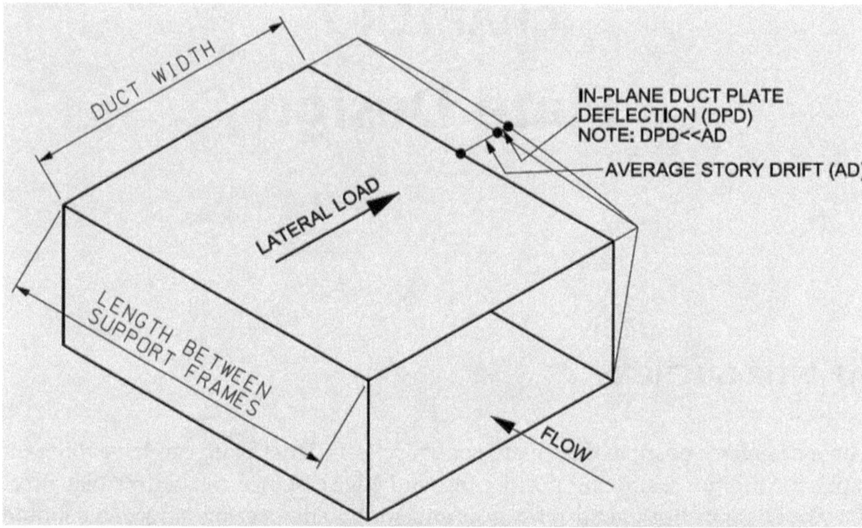

Figure 7-1. Story drift and duct plate "diaphragm" action.

to the supports based on a beam analogy. The stiffness of the duct support steel and the foundations should be included in the analysis of statically indeterminate supported ducts, as discussed in Chapters 2 and 6, to accurately determine the load path.

Planned transportation and erection modularization procedures may create loading conditions that should also be considered in the panel design. The overall duct and local stiffening arrangement necessary to serve such schemes may control some stiffener spacings and possibly the duct plate thicknesses. As an example, if a duct section must span a very long distance over a road or a building, the erection procedure may involve lifting the entire assembly in one piece, or it may be erected in two or three large pieces with temporary supports. This is similar to the methods of constructing a large-span bridge. These two procedures would impose different dead load forces in the duct and the duct plate.

7.1.2 Duct Support Scheme Effects on Plate Forces

Forces in the duct plate often depend on the stiffness of the structural steel or foundations supporting the ductwork. Figure 7-2 shows an example of a poor duct support arrangement that should be avoided. With the support conditions shown, significant torsional forces exist across the duct cross section because of the differences in support stiffness: two supports are rigid hangers, one is off a rigid column, and the fourth is near the middle of a relatively flexible beam. When distortion-induced effects like this are not addressed in the analysis and design, the actual primary load paths will *not* be those that were assumed or determined by the structural engineer in the analysis. Because of this one soft support, significant secondary forces can develop in the duct plate as the duct composite structure

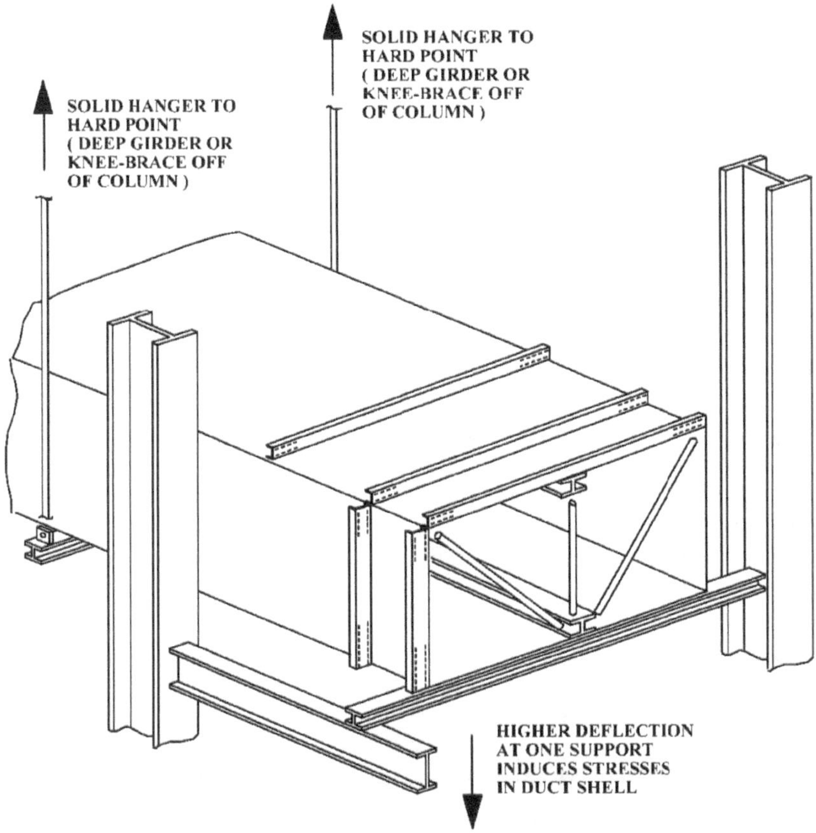

Figure 7-2. Example of poor duct support arrangement.

redistributes to carry the primary loads to the three hard supports. This behavior is possible because of the relatively high stiffness of the duct itself. Secondary forces, as described earlier, are best avoided by revising the support conditions rather than including these differential support stiffness conditions in the analysis by modeling each support with a different spring force. The effects of differential support movement might also be minimized by effective positioning of an expansion joint, as described in Chapter 2. The duct engineer may also specify maximum relative displacements of the support structure at the duct support points for each section of duct between expansion joints.

7.2 RECTANGULAR DUCTWORK PLATE DESIGN

In rectangular ducts, the plate acts in conjunction with other structural elements to balance the pressure forces and carry the gravity and lateral loads to the supports. The plate is probably the most important structural element in the duct, but it could not economically function without its stiffeners and internal trusses or moment frames.

When the duct stiffeners are in a parallel configuration, which is the usual case, a strip of plate is assumed to span and be supported by two stiffeners. The stiffeners may be oriented perpendicular or parallel to the flow, depending on the stiffener framing philosophy. Banded stiffeners oriented perpendicular to the flow are supported at the duct corners by the mating stiffener, duct plate, internal trusses, or struts. Alternatively, the stiffeners may be parallel to the flow and be supported by internal trusses, struts, or moment frames, as in Figure 7-2.

An alternative stiffener arrangement scheme is to provide multiple rows perpendicular and parallel to the flow, which is commonly referred to as waffle framing. This stiffener arrangement results in two-way bending of the duct plate and typically allows for a thinner duct plate design. The required flexural strength (M_r, as defined in the AISC's *Specification for Structural Steel Buildings*, 2017) may be determined using the methods presented in *Roark's Formulas for Stress and Strain* (Young, Budynas, and Sadegh 2011). The waffle framing concept also reduces concerns about the lateral support of the compression flange of stiffeners and may reduce the installation cost of the insulation and lagging because less sub-girts are required. However, it involves more pieces and more welding than the simple banding arrangement, so it is more labor-intensive, and this often offsets the other savings. A computer structural analysis may be needed where a complicated waffle-type grid of stiffeners is used.

7.2.1 Typical Plate Thickness and Stiffener Spacing

Duct panels for power plant ductwork are typically fabricated from steel plate with a minimum thickness of 3/16 in. (5 mm). For lighter industrial ducts, the minimum may be 10 gauge (3.5 mm) sheet. To guard against highly erosive ash or an aggressive gas environment, an erosion or corrosion allowance is sometimes considered in the plate design. The plate thickness may increase as stiffeners are eliminated as a cost reduction strategy, or it may increase because it is required to support heavy loading, such as ash load. Ductwork stiffener spacing usually will vary between 2 ft and 4 ft (0.6 m to 1.2 m) for typical plate thicknesses between 3/16 in. and 3/8 in. (5 mm and 10 mm). However, in locations of low stress and no dynamic concerns, the spacing may be 5 ft (1.5 m) or more. The stiffener spacing will depend on the loading, material, temperature, analysis method, and forces from the duct global analysis. The stiffener spacing may also be a function of practical panel shipping sizes and field splice requirements, as discussed in Chapter 10.

7.2.2 Stiffener Contributory Areas and Plate Analysis

When arranging the structural system of a duct section, the structural engineer should first develop a consistent load path that considers the relative stiffness of contiguous elements. The plate analysis and design must be based on this assumed load path arrangement. As the stiffener framing and truss arrangements vary, so will the portion of panel loads resisted by a given stiffener element. Depending on the plate and stiffener configuration, the structural engineer can assume that the duct plate spans either simply between two stiffeners or continuously over many stiffeners. In the example shown in Figure 7-3, longitudinal strips X and Y are assumed to be supported by banded panel stiffeners at Z spacing. Plate bending calculations can then be performed for a given plate strip using conventional beam theory. The load-carrying capacity of the plate then determines the plate thickness or the stiffener spacing. Additional calculations should be developed where plate stresses are applied orthogonally, such as at point 1 or point 2. At these locations,

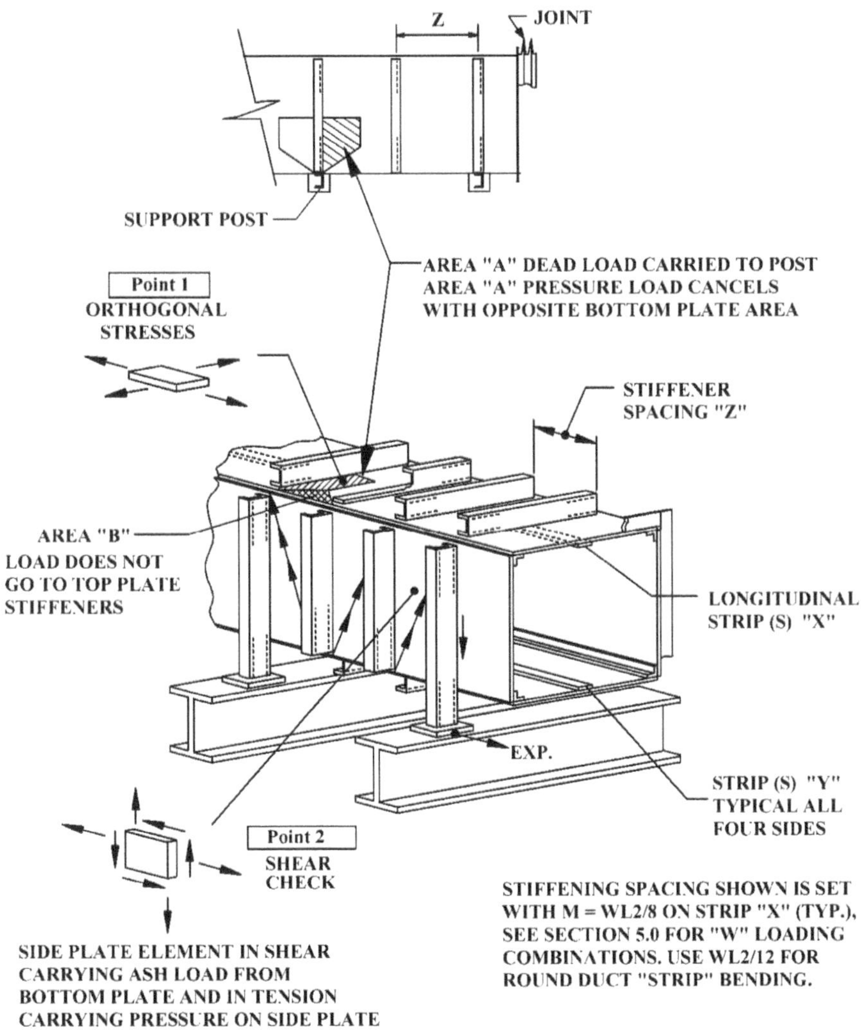

Figure 7-3. Sample rectangular duct stiffener framing.

tension field action stresses from girder shear, combined with normal stress, may greatly affect the plate and weld designs. This is discussed further in Section 6.3.2.

Many other stiffener framing patterns are encountered that are not so simple. This is especially true in transition sections of ductwork. In these unusual and difficult cases, similar assumptions can be made to simplify the analysis.

7.2.3 Plate Analysis Deflection Theories

The analysis and design of duct plate in accordance with classical flexural membrane behavior, as discussed in Section 7.2.2, is classified as small deflection theory. Small deflection theory is a simplified, conservative approach to ductwork structural analysis and design.

Large deflection theory analysis may be performed if the structural engineer wants a more detailed and economical design or is working on a retrofit project where an existing duct system must be investigated closely. Large deflection theory uses the membrane behavior of the duct plate to resist loading normal to the surface of the duct plate (Figure 7-4).

Large deflection theory better approximates reality in that bending and membrane stresses both contribute to the total stress in the plate, usually between stiffeners. The result can be up to a 20% increase in stiffener spacing or up to a 30% increase in load-carrying capacity as compared to the basic elastic strip analysis and design method. Boundary elements of the duct plate (stiffeners, corner angles, and other plates) must be capable of resisting the substantial forces developed at the boundary edges of the duct plate when the large deflection theory analysis and design method is used.

In a dynamic environment with wide stress ranges that can create fatigue problems, basic elastic small deflection theory, or strip analysis, is often used over large deflection analysis. This analysis and design method is used in a dynamic environment because additional dynamic criteria should be investigated, which typically governs over the large deflection theory design.

7.2.4 Secondary versus Primary Stresses

Both primary and secondary stresses should be considered in the duct plate design.

Primary stresses are stresses that do not redistribute or reduce after the material yields. Primary stresses are caused by loads that are independent of structural deformations. The structure will continue to deform under primary loads until failure occurs. Examples of ductwork loads that create primary plate stresses are dead load, live load, and pressure.

Forces generated by the axial deformation of metal expansion joints with thermal expansion are considered to cause primary stresses. These stresses are additive to other stresses and may be bidirectional if the metal expansion joints have been axially preset in their cold position. Also, toggle duct sections generate primary shear stresses at their joint flanges and in the adjacent duct plate. See Sections 1.4 and 2.3.2 for discussions on toggle duct sections.

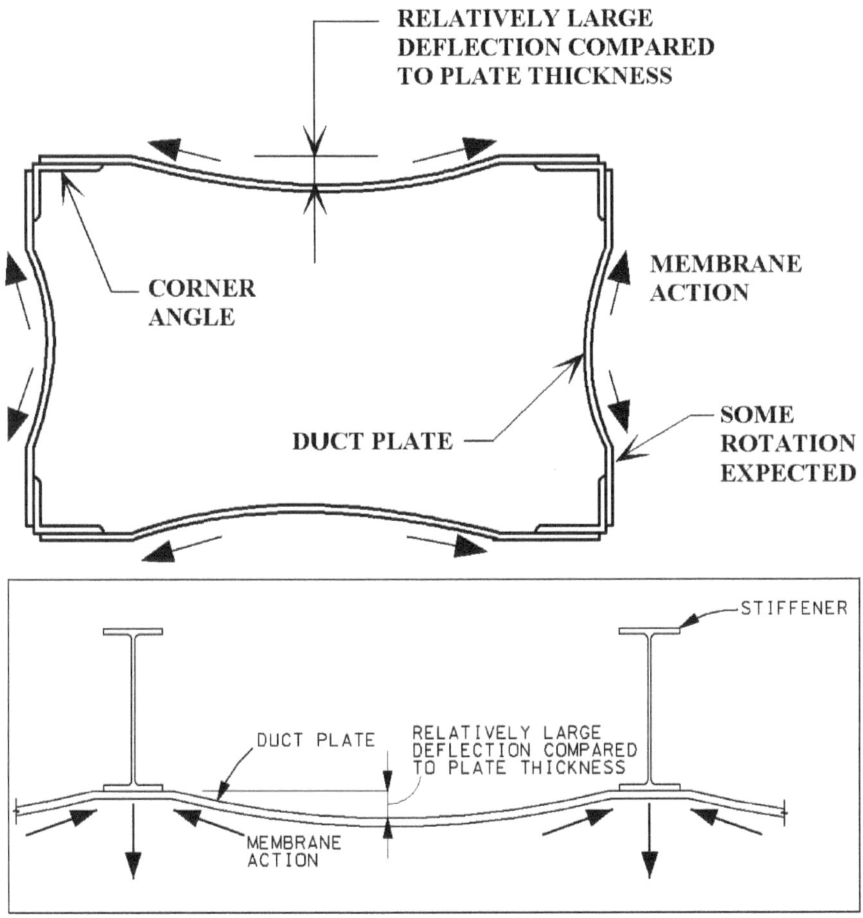

Figure 7-4. Large deflection theory behavior.

P-δ effects may need to be considered in the plate stress calculations when compressive forces are applied to the panel edges in the plane of the panel or when the global structural analysis results in longitudinal compressive stresses that are perpendicular to the local duct deflections between duct stiffeners. Global deformations in the plane of the duct plate are typically insignificant, so P-Δ effects are typically ignored in the duct plate design.

Secondary stresses are caused by self-limiting strain-controlled loads. Secondary stresses are caused by a driving force that acts only up to the point where the imposed strain ceases to increase. Examples of secondary plate stresses are those caused by constraint from the structure, stresses from friction forces caused by thermal expansion, thermal stresses from the sun or differential gas temperatures, and stresses caused by discontinuity effects at welds. Most of the time, secondary stresses are not calculated directly but are considered by the structural engineer when determining safety factors or required design margin.

Two sources of thermal secondary stresses in duct plate are cooling of welds, and thermal gradients. Cooling (and shrinking) of welds, especially continuous welds, can create residual stresses and plate warping. Thermal gradients, or differences in temperature from one area of a duct to another, can cause large plate stresses from differential thermal expansion. Depending on the exact stiffness, a large thermal gradient can cause either the hotter section of plate to buckle in compression or the cooler section of plate to tear or crack from tension, as the plate tries to achieve equilibrium or strain compatibility. These thermal gradients are often caused by deep ash deposits, gas stratification, or water intrusion. For a number of reasons, excursion temperature distribution within a duct section, and thus the thermal gradient, may be very different from operating temperature distribution.

Thermal loadings should be considered in certain applications, such as in large structures or where temperature differences are large enough to cause significant stresses. Thermal stresses can easily dwarf primary stresses, so the structural engineer should first try to minimize the thermal effects either through process change, including adding flow control and mixing devices, or by rearranging the stiffness of the duct structural system so that secondary stresses will be minimized.

The duct element framing arrangement should generally be as flexible as possible in a strategic effort to minimize thermal plate and member stresses. In cases where the secondary stresses are relatively small, consideration can also take the form of conservatively limiting the primary stress levels by either increasing the factor of safety or by using conservative operating loads.

The structural engineer should be aware that where high secondary stresses from shrinkage or thermal loadings are known and unavoidable, the ASME *Boiler and Pressure Vessel Code, Section VIII: Rules for Construction of Pressure Vessels* (2015c, Part UG-23) recommends that the combination of all primary and localized secondary stresses over the full stress range should be limited to twice the material yield strength, reduced for higher temperature, to ensure shakedown of the overall structure to elastic action. Secondary stresses should be combined with the primary based on the criteria in Section 7.2.5.

For ducts with operating temperatures above 150 °F (65 °C), dissimilar metals with significant differences in coefficient of thermal expansion should not be welded together, especially if the pieces are very large and buckling could easily occur. If welding must be performed, the structural engineer should ensure that the coefficients of expansion of the dissimilar metals are not substantially different. For example, welding carbon steel plate to austenitic stainless steel plate can introduce very large thermal stress in hot situations.

7.2.5 Combined Local and Global Interaction

The structural engineer should be careful to consider locations where a plate element is stressed in two orthogonal directions. An example of this condition is the plate section that is working compositely with the stiffener and also in bending, like point 1 in Figure 7-3. Another example of combined stress is when the panel plate acts as a global structural element in transmitting gravity or wind loads to the duct supports, like point 2 in Figure 7-3. The plate has the mentioned two directions of bending stress plus shear stress, as the panel acts as a girder or truss to distribute the loads to the supports.

The multiple states of stress resulting from the duct plate's participation in multiple structural systems from all load combinations should be properly combined. The structural engineer should choose the stress or force combination techniques that are consistent with the structural analysis techniques used for the duct plate and the global duct structure. The duct panel design should consider the stresses or forces resulting from the structural global analysis in addition to the local bending. Tensile, compressive, and shear stresses or forces should be combined using a well-established method.

The duct plate should be evaluated for combined stresses or forces as illustrated in the AISC *Specification for Structural Steel Buildings* (2017). The available flexural strength (M_c) of the duct plate may be determined based on the specification's provisions for rectangular bars bent about the minor axis. Therefore, the lateral-torsional buckling limit state does not need to be considered. Similarly, the available flexural stress (F_{cbw}) may be determined based on the specification's provisions (in Section H2) for unsymmetrical shapes subject to flexure and axial load. The lateral-torsional buckling stress will not control the yielding limit state and does not need to be considered.

Alternative stress combination techniques are presented in various structural engineering textbooks. Some that are generally used by structural engineers designing ductwork include direct algebraic addition:

$$f_a + f_b < F_{all}$$

and development of principal stresses:

$$\sigma_1 < F_{all}$$

$$\tau_{max} < F_v$$

where

f_a = Calculated axial stress,
f_b = Calculated bending stress,
σ_1 = Calculated principal normal stress,
τ_{max} = Calculated principal shear stress,
F_v = Allowable or design shear stress, and
F_{all} = Allowable or design normal stress.

7.3 CIRCULAR DUCTWORK PLATE DESIGN

Circular ductwork is typically analyzed and designed based on the theory of thin-walled shells, given the very large diameter to wall thickness (D/t) ratio. Depending on the D/t ratio, the plate may be strong enough to act without stiffeners, like a large pipe, or the plate may require stiffening or internal bracing, like a rectangular duct. The plate is the most important structural element in circular ductwork.

Additional guidance on the design of round ductwork can be found in STS-1 *Steel Stacks* (ASME 2011), *Boiler and Pressure Vessel Code, Section VIII* (ASME 2015c), and Troitsky's *Tubular Steel Structures Theory and Design* (1991).

7.3.1 Designing for Buckling

Understanding overall and local buckling behavior is imperative in the design of round ductwork. Thin-walled shells under compressive longitudinal stress may buckle either as a whole about the longitudinal axis (Euler column buckling), or locally in the wall. Local buckling strength depends on the ratio of the plate thickness to the diameter. Thus maximum diameter-to-thickness ratios (D/t) are typically used to evaluate and restrict buckling. For reference, a D/t ratio of approximately 300 is typical in large pipe design where the pipe has low negative pressures. Shell buckling may control the plate design for ductwork with a D/t ratio between 300 and 500, but when D/t exceeds 500 then it is likely that multiple bands of stiffeners will be needed, especially with a negative pressure loading. The structural engineer should consider that designing, detailing, and fabricating multiple bands of stiffeners may be more expensive than simply increasing the plate thickness.

Round ductwork is typically assumed to act as a beam-column with a cross section equal to the entire duct tube. To ensure this behavior, the duct tube cross section needs to remain essentially circular. It should be taken into account that round ductwork under bending will tend to flatten or oval, lowering its flexural stiffness and strength. This is commonly referred to as the Brazier effect. The longitudinal compression and tension stresses caused by the applied bending moment are directed inwards toward the center of the circular cross section. As these stresses increase, the structure becomes unstable and locally collapses/buckles to form a kink. Stiffeners should be spaced as necessary to prevent any flattening or ovaling of the duct plate so that the local buckling strength can be determined based on thin-shell principles without consideration of the ovaling effects. A stiffener spacing equal to the duct diameter is recommended, based on past industry practice, but in no case should the stiffener spacing or the duct plate thickness be less than the requirements specified in ASME STS-1 (2011). Where stiffeners are not provided to prevent ovaling, the critical buckling stress should be determined with consideration of the ovaling deformation of the ductwork.

7.3.2 Longitudinal Stress

The duct loads discussed in Chapter 4 develop moments and axial forces on the duct tube cross section that result in longitudinal (or meridional) stresses parallel to the axis of the duct cylinder. This stress is analogous to the outer fiber stress in conventional beam bending design (Figure 7-6). This stress is typically minimal for circular ducts with a small span-to-diameter ratio (L/D). A ratio of 10 or less is recommended, because circular duct sections with L/D ratios above 10 may develop significant longitudinal stresses.

Based on the selection of the support configuration, the duct plate design should address circumferential bending stresses if the supports are positioned normal to the duct surface, or concentrated tangential shear stresses adjacent to the supports if the supports are oriented tangentially to the duct plate. The introduction of a stiffener ring at the support points to help resist the localized hoop stresses and collect forces for transmission to the supports will induce meridional bending stresses in the duct plate. This meridional bending stress is caused by the change in the radial stiffness of the duct and the stiffener ring at the supports. The interaction of longitudinal/meridional stress, hoop stress, and the tangential shear stress in the support region should be addressed by the structural engineer. Guidance on the combination of stresses for the multiple design conditions is given in Section 7.2.5.

Figures 7-5 and 7-6 show the multiple concurrent forces or stresses that should be resolved through the duct shell to the supports. In Figure 7-6, a static

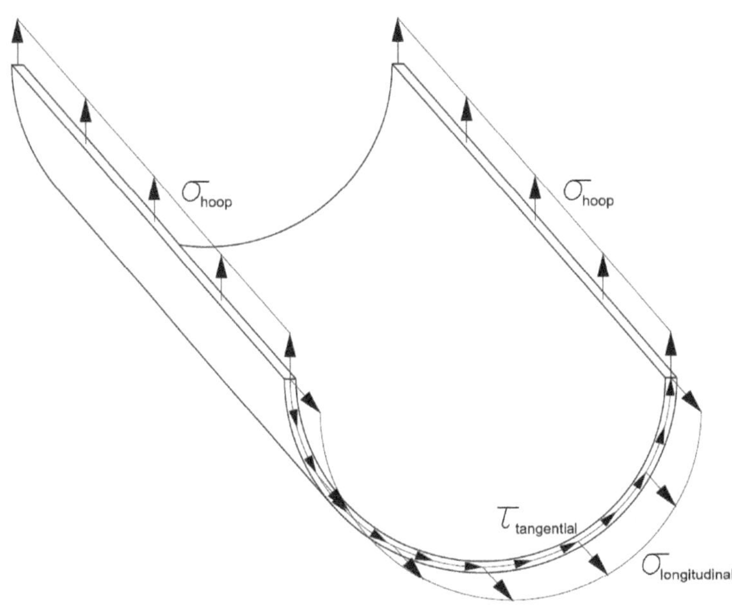

Figure 7-5. Circular duct plate stresses.

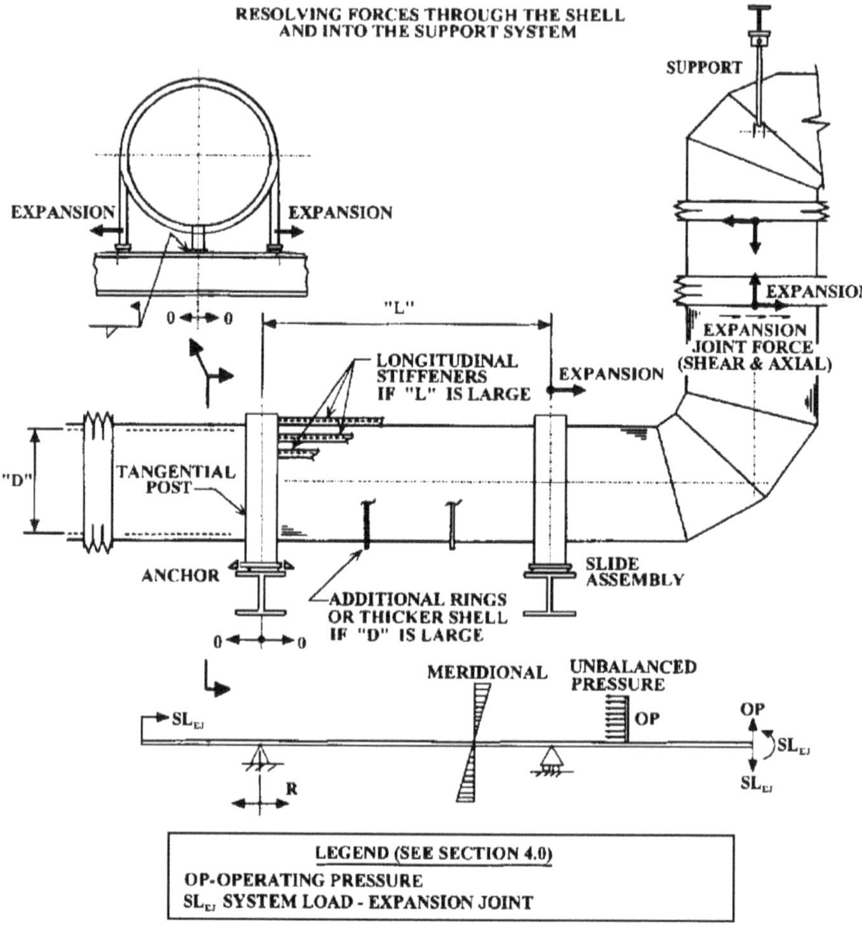

Figure 7-6. Example of circular duct loading considerations.

analysis reveals overturning moments from unbalanced pressure and expansion joint forces, which lead to the undesirable potential for uplift.

The longitudinal stresses caused by major-axis and minor-axis bending of the duct tube are amplified by several orders of magnitude at the duct elbows. These secondary stresses are a result of the tendency of the duct tube to oval from the greater flexibility in the elbow regions. Duct elbows should be detailed accordingly, and stiffeners should be added to reduce the flexibility. Typically, duct stiffeners are placed at both ends of the elbow and along the mitered sections to provide sufficient rigidity to ensure that the duct tube remains circular, to prevent the secondary amplification longitudinal bending stresses from developing.

The American Iron and Steel Institute's *North American Specification for the Design of Cold-Formed Steel Structural Members* (2012) provides guidance on the capacity of thicker-walled ($\frac{D}{t} \leq 0.441\frac{E}{F_y}$) cylindrical tubular structures. Round ductwork typically exceeds the specification's limitations, and therefore proven structural engineering principles such as those provided by ASME STS-1 and Troitsky should be used to determine the axial and flexural capacity.

7.3.3 Shell Hoop Stress

Hoop stresses are usually a minor contributor to the overall stress, even at high pressures. This is one of the major advantages of circular ducts, because stiffeners and their associated costs are not usually needed for pressure loading stresses, unlike for a similarly sized rectangular duct. The hoop stress from an internal pressure p is $\sigma = \frac{pr}{t}$, where r is the mean radius of the duct tube and t is the duct plate thickness. The hoop stress in a duct elbow section is smaller or larger than that of a straight section, depending on the change of curvature. A maximum hoop stress, $\sigma_{max} = \frac{pr}{t}\left(\frac{R-\frac{r}{3}}{R-r}\right)$ (Troitsky, Equation 9.6), occurs along the inner bend radius (Figure 7-7).

In addition, local longitudinal stresses are developed in a duct elbow from the uniform internal pressure load and should be combined with the global longitudinal stresses.

For large circular ducts, the distribution of wind pressure about the circumference should be considered. The nonuniform distribution of normal pressure loads about the duct will induce additional circumferential bending stresses in the plate, which are additive to the hoop stresses. Guidance on the distribution of wind pressures about the duct's circumference can be obtained from ASCE Task

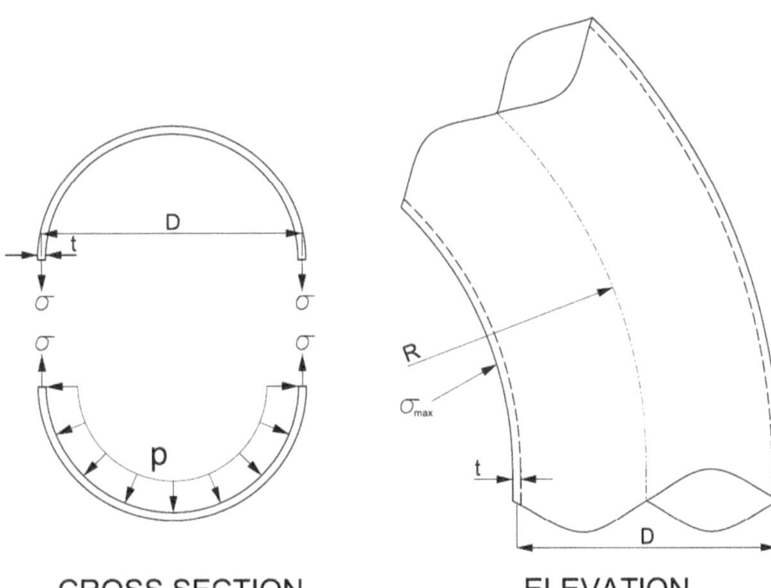

Figure 7-7. Shell hoop stress.

Committee on Wind Forces (1961). Clarification on determining the circumferential stress from wind load is provided by STS-1 (ASME 2011).

Hoop stresses may also be developed by the interaction between the duct support system and the duct shell at the duct supports. These hoop stresses should be combined with the local hoop stresses caused by the internal pressure. The magnitude of this hoop stress is a function of the location of the support points along the duct's perimeter.

7.3.4 Tangential Shear Stress

When bottom support posts or ring girders are aligned tangentially with the side of a circular duct, as shown in Figure 7-6, the vertical gravity loads are carried by the duct plate as tangential shears to the supports. The distribution of this tangential shear is a function of the location of the supports about the duct circumference and whether the duct is stiffened at the support. Locating the supports in this manner minimizes the introduction of circumferential bending stresses within the duct shell at the supports and maximizes the overturning resistance to wind and seismic loading perpendicular to the duct run.

The tangential shear stresses caused by the torsional moments produced at duct elbows should also be combined with the global vertical and horizontal shear forces as applicable to determine the total design tangential shear stress. An example of combining these stresses is provided in Troitsky (1991), Figure 7-8.

Evaluation of tangential shear through the duct section may be determined from a finite element structural analysis or by application of prudent analytical techniques based on the type of support configuration used. Zick (1971) *Stresses in Large Cylindrical Horizontal Vessels on Two Saddle Supports* and Troitsky provide recognized approaches for evaluation of stresses at support regions. The approach presented in Zick's paper is consistent with the recommendations in ASME (2015c, Section 4.15). Based on this latter criterion, tangential shear stress is limited to 80% of the allowable stress (S) as defined in ASME (2015c).

7.3.5 Factors of Safety

The corresponding AISC or ASME factors of safety are typically applied, depending on the code basis used for round ductwork design. The design code and applicable safety factors should be selected by the structural engineer to ensure that the design is within the limits of the code basis. For example, the AISC specification does not address the buckling behavior of thin-walled shell structures, so a factor of safety of 2 to 3 is typically selected for buckling in the duct shell, depending on the type of loading, either from external pressure or axial compression, consistent with ASME (2015b, Part D, Mandatory Appendix 3 3-600). However, the structural engineer may increase the safety factor if the quality of the fabrication and erection is questionable.

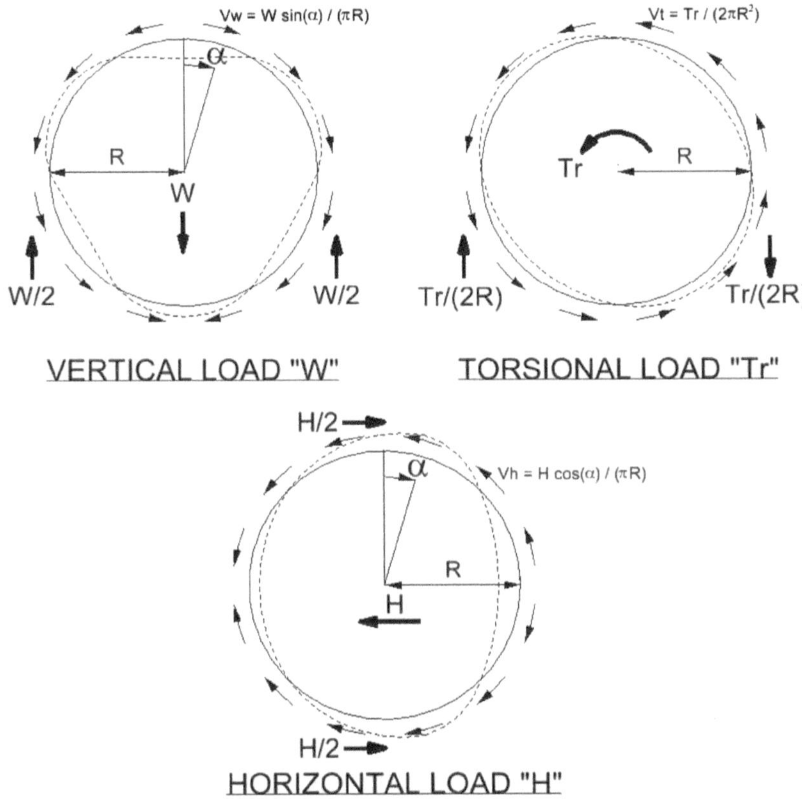

Figure 7-8. Tangential shear stress and moment diagram.

7.3.6 Material Handling Considerations

Circular ducts have a special sensitivity to damage in transportation and handling because of the potential for buckling and ovaling. If a duct section is to be shipped in one piece as a cylinder, it should be analyzed and designed for impact during shipping and handling. This would include the supports, bracing, internal elements, and the plate. See Chapter 10 for more information on handling and for a discussion on bracing for shipping. A construction handling impact factor of 1.5 is considered appropriate. There are no established rules for a handling criterion that may control the plate thickness. A 3/16 in. (5 mm) minimum thickness is often used for diameters up to 48 in. (1.2 m), and a 5/16 in. (8 mm) minimum thickness is common for diameters up to 90 in. (2.3 m) where stiffeners are not provided.

7.4 OTHER CONSIDERATIONS

There are additional concerns other than stress and vibration that need to be considered by the structural engineer when sizing duct plate, although stress is a contributing factor in some of these. The most important of these concerns is crack control.

7.4.1 Fatigue, Endurance Limits, and Crack Control

Although more applicable to the design of metal expansion joints, the endurance limit concept also applies to duct plate in the area of a large fan or other source of pulsating signal or forcing function. Mechanical cycling of the plate, whether induced by pressure pulsations, sonic acoustically induced vibrations, or resonance, can eventually cause a fatigue failure. The four primary parameters for fatigue are the number of stress cycles, the stress range, the magnitude of stress concentrations, and the construction details. To establish a reasonable cycle life, the plate stress is typically kept low enough that fatigue failure will not occur. This allowable stress level is called the endurance limit. Low cycle / high stress fatigue situations should always be avoided.

Design codes address fatigue in various ways. For example, the ASME *Boiler and Pressure Vessel Code, Section I: Rules for Construction of Power Boilers* (2015a), does not address thermal fatigue failure analysis directly; it only addresses fatigue by conservatively involving the use of the Rankine maximum principle stress theory. The AISC specification uses category factors for various typical construction details to address mechanical fatigue. The International Committee on Industrial Chimneys *Model Code for Steel Chimneys* (CICIND 2010) and ASME (2011) give examples for the specific situation involving vortex shedding external to circular chimneys and steel stacks. For boiler pressure parts, some duct and boiler engineering companies use maximum-thickness rules to limit cyclical thermal shock cracking caused by restraint.

Fan outlet ductwork is particularly sensitive to fatigue from internal vortex shedding and vibrations caused by flow pulsations from the fan blades. The Air Movement and Control Association's *Fans and Systems* (2007) describes how diffusion from the fan is controlled by specific geometric angles such that a uniform velocity profile is established 3 to 5 duct diameters downstream of the fan. The use of this criterion should minimize pressure drop and improve flow conditions and may reduce fan power consumption.

Although fatigue damage is closely associated with operation procedures and design conditions, a measure of crack control can be obtained in duct plate, and generally in all structures, by the implementation of the following considerations.

- The natural frequency of all ductwork structural elements should be tuned away from the fan's pulsation frequencies by a reasonable factor, such as 20%. This may be achieved by adjusting the stiffener spacing, the element length, the element stiffness, or the connection fixity.
- Plate stresses should be low, and stress ranges should be minimized.
- The plate material should have adequate toughness. The acceptable toughness should be determined by the structural engineer based on the material's lowest operating temperature. Charpy V-notch impact tests and minimum toughness requirements can be specified for the material if required.

- Construction details should minimize stress risers, especially at connections.
- Plate material flaw sizes should be controlled with proper welding, fabrication, inspection, and erection.
- Multiple redundant load paths should be established in the structure. Cracks tend to arrest themselves as the structure redistributes the stresses.
- The loading rate should be controlled, if possible.

7.4.2 Solutions to Problems

Three considerations will help produce an acceptable duct plate design in a dynamic environment: applying experience, developing and using empirical design methods, and providing field adjustments.

Applying experience. The structural engineer should interact with the mechanical process engineer and/or the customer before the start of the design to avoid conditions that could shorten the duct's design life. For example,

- Large fans may operate with large pressure pulsations caused by poor inlet or outlet configurations.
- Large ducts that require large vortex shedding internals can develop standing sonic waves and pulsations.
- Ducts with a high flow velocity can induce vortex shedding.

Many ductwork system arrangements are too complex to realistically model mathematically and accurately analyze. Keeping the duct configurations as simple as possible should be a generic strategy during the conceptual and layout phases of the work.

Empirical design methods. With experience and by extensive analysis of historical data, analysis methods, and design rules, empirical equations can be established to calculate the dynamic response of the duct plate to excitations both circumferentially and longitudinally. Equations defining permissible deflections and identifying natural frequencies can be found for some simple configurations. Tuning the elements' natural frequencies 20% or more away from expected signal frequencies should avoid resonance, rapid cycling, and fatigue cracking. If this is properly done, then a reasonable cycle life should be attained. There is more discussion of this topic in Chapters 8 and 9.

Field adjustments. For sonic and resonance problems in particular, the solution is sometimes performed reactively. Following the collection of field data, minor changes can be made to significantly change natural frequencies, reduce deflections, and reduce stress ranges. Examples of changes are reframing the stiffeners in an area, revising internal truss arrangements, increasing plate thickness, reinforcing elements, adding a corner angle, and repairing damage by drilling crack ends and patching. All of these modifications change the stiffness of the individual elements and the overall duct structure. If possible, the best solution is to eliminate the source of the forcing function.

References

Air Movement and Control Association. 2007. *Fans and systems: Publication 201-02 (R2007)*. Arlington Heights, IL: Air Movement and Control Association.

AISC. 2017. *Specification for structural steel buildings*. AISC 360-16. Chicago: AISC.

AISI (American Iron and Steel Institute). 2012. *North American specification for the design of cold-formed steel structural members: S100-12*. Washington, DC: AISI.

ASCE. 2017. *Minimum design loads and associated criteria for building and other structures*. ASCE 7-16. Reston, VA: ASCE.

ASCE Task Committee on Wind Forces. 1961. *Wind forces on structures: ASCE transactions 126*, 1124–1198. Reston, VA: ASCE.

ASME. 2011. *Steel stacks*. ASME STS-1. New York: ASME.

ASME. 2015a. *ASME boiler and pressure vessel code, Section I: Rules for construction of power boilers*. New York: ASME.

ASME. 2015b. *ASME boiler and pressure vessel code, Section II: Materials*. New York: ASME.

ASME. 2015c. *ASME boiler and pressure vessel code, Section VIII: Rules for construction of pressure vessels, Division 1*. New York: ASME.

CICIND (International Committee on Industrial Chimneys). 2010. *Model code for steel chimneys*. Brighton, UK: CICIND.

Troitsky, M. S. 1991. *Tubular steel structures: Theory and design*. Washington, DC: James F. Lincoln Arc Welding Foundation.

Young, W. C., R. G. Budynas, and A. M. Sadegh. 2011. *Roark's formulas for stress and strain*. New York: McGraw Hill.

Zick, L. P. 1971. "Stresses in large cylindrical horizontal vessels on two saddle supports." Welding Journal Research Supplement. Accessed April 6, 2020. http://fliphtml5.com/auhu/exbm/basic.

CHAPTER 8
Structural Element Design

8.1 GENERAL CONSIDERATIONS

The overall structural integrity of a duct can only be as reliable as its individual elements. The structural elements must perform their duty across a range of operating and environmental conditions for the design life of the duct. To meet this demand, the overall duct structure and each individual structural element should be designed by the structural engineer to meet these conditions while remaining within certain predefined safety factors. To properly design the individual structural elements, the structural engineer must be cognizant of the many different failure modes an element or structure can experience. Once these failure modes are recognized, the structural engineer should investigate the different types of loading conditions the elements will face during the life of the duct. These load combinations are presented in Chapter 5.

In general, ducts may be quite large. Structural engineers can take advantage of the inherent strength of these ducts by adequately reinforcing the plate material to withstand not only internal pressure loads but also membrane and bending stresses from dead load, live load, and other operating and environmental loads. The design of structural elements entails identifying the different forces on the duct and combining their effects using prudent structural mechanics so that the resultant may be compared to the applicable structural strength of the element. This process may involve complex structural analyses and may be performed by hand or with the aid of computers. Either way, the structural engineer must be aware of the various, and sometimes redundant, paths by which loads can get to the external restraints.

When performing the structural design of a duct and its various structural elements, the structural engineer's duty is to provide adequate structural integrity and to design the structural elements economically. The loads must be safely transferred from the duct down to the foundations or, in the case of most pressure and friction loads, cancelled out within the structure. An economical structural system is also usually less rigid, so it has lower strain-controlling stresses such as those developed from thermal expansion and temperature differentials.

Ducts are inherently indeterminate structures with complex loading patterns imposed on them, so it may be difficult to intuitively recognize the load paths that will adequately transfer the loads to the supports. A structural engineer must be able to identify the various load paths and properly design each element along the paths.

There are two basic duct configurations: rectangular (Section 8.2) and circular (Section 8.3). Of course there are many other possible shapes, especially where duct sections transition into major equipment. Unusual duct configurations are not specifically addressed here because of the infinite number of possible shapes. However, once the design of rectangular and circular duct sections is mastered, the structural engineer should be able to use judgment and similar principles when designing unusual duct configurations.

8.2 RECTANGULAR DUCTS

8.2.1 Stiffener Layout

Stiffeners are attached to the duct plate to strengthen the flat panels against pressure loads, to reduce vibration of the plate, and to stiffen the plate against buckling. If a duct's panel height or width is greater than the allowable plate span for the design loads or for vibration considerations, then stiffeners are required to reduce the span. Stiffeners should be spaced at some convenient spacing so that the plate has adequate strength. The design of duct plate is discussed in Chapter 7. Stiffeners are usually placed to make uniform panels. If the panel height or width is less than the allowable plate span, then theoretically the panel does not need stiffeners. However, if the duct is to be lifted and handled in one piece, then some stiffening may be necessary to maintain the duct's shape. In any case, stiffeners become an integral part of the duct's structural system. Stiffeners may also function as duct supports, as well as plate reinforcement.

Placement and Spacing. Stiffeners are most often located on the outside of the duct. This avoids disturbing the flow and contributing to pressure drop. They may also be located inside the duct, but the effect on pressure drop, erosion, and corrosion should be considered. There may be many acceptable stiffener arrangements, but the most economical are the ones that make use of stiffeners in more than one capacity. This limits the number of pieces and the amount of welding. External stiffeners are usually oriented transverse to the flow, spanning from one edge of the panel to the other and effectively wrapping around the duct. Internal stiffeners are usually oriented parallel to the flow, spanning between internal trusses, moment frames, struts, or perpendicular collector stiffeners. The overlap between an internal and external stiffener should be detailed accordingly if outward loads are to be transferred between the stiffeners. The duct plate is typically not adequate to transfer the outward loads, so a more robust connection is required. Inward loads may be transferred through bearing action between internal and external stiffeners.

Long runs of uninterrupted duct usually have stiffeners arranged in an orderly and uniform manner. If possible, the distances between stiffeners should be equal, even if it means adding an extra stiffener or two. Equal spacing results in more uniform stresses in both the plate and the stiffeners, as well as simplifying fabrication and the installation of insulation and lagging. An exception to this approach would be near large fans, where unequal plate spans would help change the duct's fundamental natural frequency, reducing plate vibrations.

Placement of stiffeners around structural discontinuities, such as access doors and flue-gas sampling probes, may be required, depending on the location and size of the discontinuity. This should be handled on a case-by-case basis to ensure that the discontinuities do not affect the overall structural behavior or develop localized overstress.

Even though the arrangements of duct systems vary greatly, they are often made up of configurations that are similar from one run of duct to another. Transition sections, rounded corners, and T-sections are basic geometries, except for dimensional differences. A standard set of preferred stiffener layouts should be developed so that a consistent approach is taken for all the duct sections.

There can be many acceptable stiffener arrangements for a given section of duct. Structural engineers tend to arrange stiffeners perpendicular to the run of the duct, but there are times when it is necessary to orient them differently. Figure 8-1 is an example of how stiffeners can be framed to form a successful design. The two

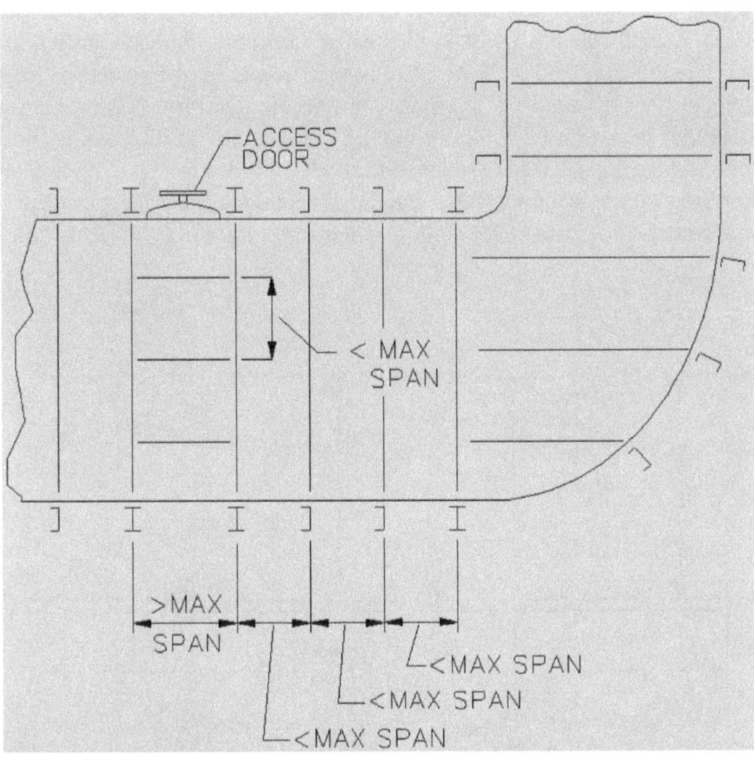

Figure 8-1. Example stiffener arrangement.

main concepts when arranging stiffeners are (1) the plate span between stiffeners must be less than the allowable span, and (2) the ends of each stiffener must be adequately attached to a stable load-carrying element, such as another stiffener or a corner angle. In some situations, such as at duct bends, it may be necessary to locate stiffeners to ensure continuity and sufficient transfer of the loads.

Stiffener Sizing Considerations. When choosing stiffener sizes, the structural engineer will usually choose as few depths and sizes as possible. Common shapes should always be used. This approach is more economical from a material standpoint and simplifies installation of insulation and lagging. The structural engineer should predetermine a convenient set of structural steel sections that includes angles, channels, and wide flange shapes to be used on a particular project.

8.2.2 Stiffener Design

Composite Action. Once a duct stiffener is welded to the duct plate, the most realistic assumption is that they act together in composite action in carrying the applied loads. Composite action is the interaction of a stiffener element and a certain portion of the attached duct plate to act as one cross section to resist loads. The duct plate width that may be considered to act compositely with the duct stiffener is called the effective width. The position of the composite section's neutral axis reflects this continuity, and it can be calculated using basic applied mechanics principles. Figure 8-2 illustrates typical composite stiffener cross sections. The effective width of the plate is determined by the structural engineer, and the calculation method may be somewhat arbitrary. There are several rationales that can be used to choose an effective width of duct plate. Blodgett's *Design of Welded Structures* (1976) recommends 12 times the plate thickness, t, off to either side of the stiffener edge. The United States Steel Corporation's *Steel Design Manual* (1981) uses $190t/\sqrt{F_y}$, which is $32t$ for $F_y = 36$ ksi: $16t$ on either

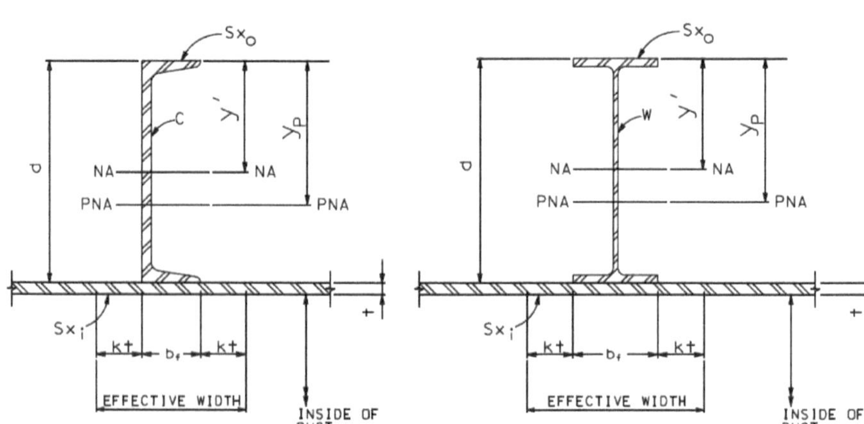

Figure 8-2. Typical stiffener composite sections.

side of the stiffener leg. The AISC *Specification for Structural Steel Buildings* (2017a) indicates that the compact flange limiting slenderness parameter (λ_p) for flange cover plates between a line of fasteners or welds subject to flexure is $1.12t\sqrt{\frac{E}{F_y}}$. This equates to $32t$ for $F_y = 36$ ksi and $27t$ for $F_y = 50$ ksi. The buckling mode of duct plate overhanging the stiffener flange edge differs from the buckling mode of a flange cover plate. The former is typically idealized based on a fixed-free configuration. The latter is similar to a pinned-pinned configuration when the cover plate has a width less than the beam flange. The former is half that of a flange cover plate, so the width-thickness ratio would be equivalent to $16t$ and $13.5t$ for 36 ksi and 50 ksi steel, respectively. In reality the duct plate is continuous, and there is some rotational restraint at the "free" end. Therefore, the actual width-thickness ratio limit to develop a fully plastic stress distribution is greater than the value based on the idealized fixed-free configuration for duct plate overhanging a flange edge.

It should also be noted that as the temperature increases, the yield strength F_y decreases faster than the modulus of elasticity E, so as the temperature increases, $\sqrt{\frac{E}{F_y}}$ will increase as well. The ambient-temperature values of F_y and E are typically used because the design operating temperature is likely to be higher than the true operating temperature, and certain loading conditions, such as dead plus wind, could occur when the duct is at ambient temperature.

The effective width should be selected by the structural engineer based on reliable publications, such as those mentioned earlier, or successful project experience. The effective duct plate width selected should consider local buckling and strain compatibility. Regardless of which criterion is used to determine the effective width, adjacent effective plate widths should not overlap. In Figure 8-2, kt is the width on either side of the stiffener.

For composite action to be fully effective, the attachment between the stiffener and the plate must be strong enough to transfer horizontal shear between the two, as calculated by the required load in the weld $f_w = \frac{VQ}{I}$, where V is the shear force, Q is the statical moment of the duct plate about the composite section centroid, and I is the composite section's moment of inertia. The most typical and effective connection is intermittent fillet welds that are staggered on either side of the stiffener. Intermittent welding adds less heat to the plate, reducing the amount of plate warping, and is more economical. This weld is typically continuous for 3 in. to 12 in. (8 mm to 30 mm) on both sides at both ends of the stiffener, and intermittent along the stiffener's length. If there are any discontinuities along the stiffener, the weld is usually made continuous around the discontinuity. The size and spacing of the intermittent fillet welds should be designed for the horizontal shear forces f_w. Figure 8-3 shows typical stiffener-to-plate welding.

There could be additional forces from the global panel analysis that also must be included when designing these welds. The minimum intermittent welding requirements presented in the American Welding Society's *Structural Welding Code, Steel* (2015) are meant to preclude local buckling of the plate to maintain the plate/stiffener composite action. In a corrosive or dynamic environment, the

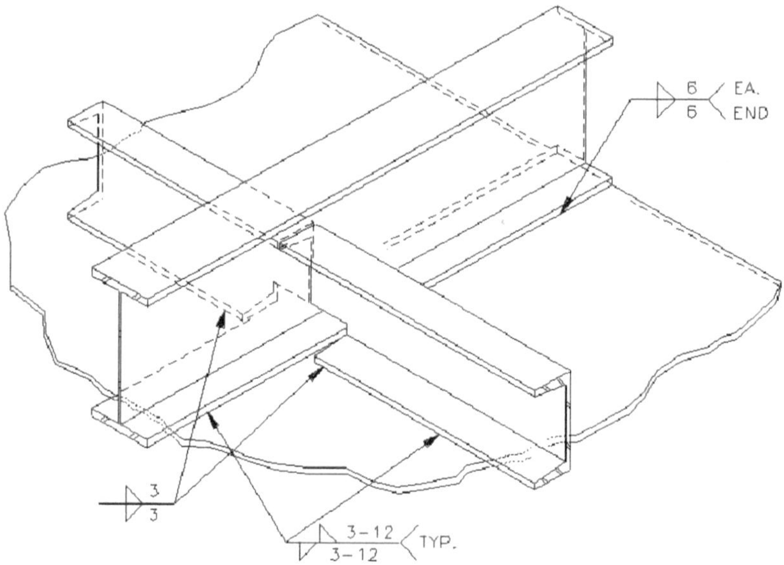

Figure 8-3. Typical stiffener-to-plate connections.

structural engineer should consider the cost-effectiveness of replacing intermittent welds with continuous welds to achieve a more reasonable design life.

Stiffener Loading. The two main categories of loading information are load types and loading direction. Loading direction controls whether the load puts a particular flange in compression or tension. Each stiffener on the duct should be checked for the effects of the various different load combinations as presented in Chapter 5. Special attention should be given to the load direction when determining the appropriate load combinations, because compression/buckling and tension limit states may need to be considered in the design of the same flange. For example, at a typical floor stiffener, the negative/suction pressure load opposes the dead and ash loads. Therefore, a load combination that includes only the negative pressure load and the resisting dead load should be considered in addition to a condition where dead, ash, and positive pressure loads are applied, because the foremost load combination results in compression of the outboard flange, but the subsequent load combination results in tension of the outboard flange (Figure 8-4). This is different from a roof stiffener, where the dead, roof live, and negative-pressure load all act in the same direction, and only compression of the outboard flange occurs for this particular loading condition. The stiffener design should consider all applicable compression and tension limit states under both the negative and positive design pressures.

Design Limit States. When composite action of the stiffener and the duct plate is not considered, then the stiffener design should consider the applicable limit states for the stiffener cross section alone, as shown in Section F of the AISC specification (2017a). In this situation the weld between the stiffener and the duct plate does not need to consider the horizontal shear transfer.

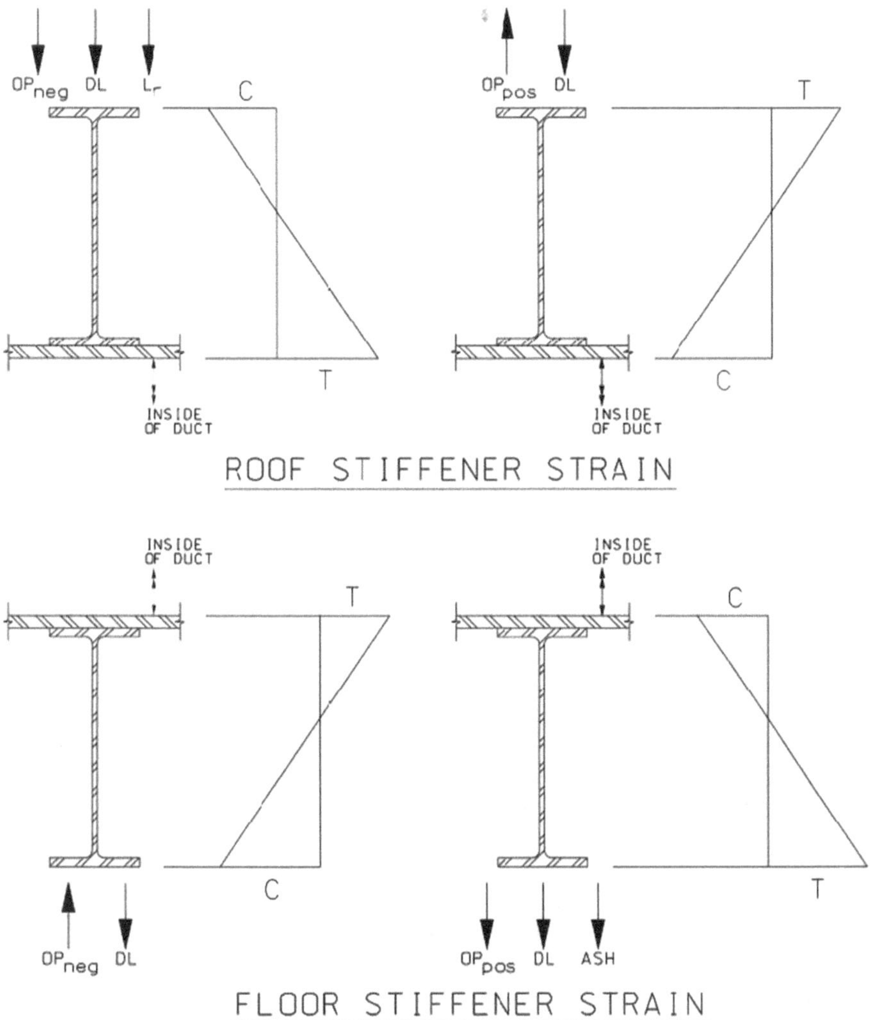

Figure 8-4. Stiffener loading and strain relationship.

A more economical design can be achieved by considering the composite action, but the resulting cross section will be asymmetrical, and the limit states used in the design should account for this asymmetrical behavior. For I-shaped members or channels, the limit states of compression flange yielding, lateral-torsional buckling, compression flange local buckling, and tension flange yielding should be considered, as shown in AISC Section F4 or F5. For other stiffener cross sections, additional limit states may need to be considered, such as web local buckling. The engineer should use prudent judgment when calculating the composite section moment strength for geometries not specifically addressed in the AISC specification, and the engineer should ensure that all applicable limit states are addressed.

The lateral-torsional buckling limit state may not need to be considered as long as there is a positive connection between the duct plate and the stiffener and the duct plate has sufficient strength and stiffness to effectively brace the flange in question. Guidance on this latter condition is provided in the AISC specification, Appendix 6, "Member Stability Bracing." It addresses both the beam lateral-torsional buckling and column bracing criteria. In some situations a strap or stiffener plate is added to the outboard flange to provide a positive means of bracing this flange (Figure 8-5).

Duct stiffeners supported at the duct corners should account for the axial load from the end reactions of the support stiffener (Figure 8-6). This axial load should be combined with the strut component due to the tension field action, if applicable, from the overall panel analysis effects that are discussed in Chapters 6 and 7. The total axial load effect should be combined with the effect from the applied moment in accordance with the AISC specification or other prudent combination techniques presented in various structural engineering textbooks.

Precalculated tables of composite cross-sectional properties and specially designed computer programs or spreadsheets can help calculate the member capacities for various loading combinations.

As a final check, the design of some stiffeners may include a deflection calculation to verify that the stiffener meets a specified deflection criterion. Deflection checks are important for horizontal panel stiffeners where water ponding or high ash loads are possible. A deflection criterion may also be used to increase the stiffener's moment of inertia to counter excessive vibration.

Stability Considerations. The AISC specification requires stability to be addressed in the analysis when member capacities are determined based on the specification's provisions. As discussed in Chapter 7, the in-plane deformations of the duct plate are typically minimal and are typically neglected in ductwork design. Therefore, P-Δ effects at stiffeners that are not part of the support frame are typically neglected, and only P-δ effects are addressed.

Ductwork support frames are similar to braced or moment frames in a building in that the lateral and/or notional loads develop global and local

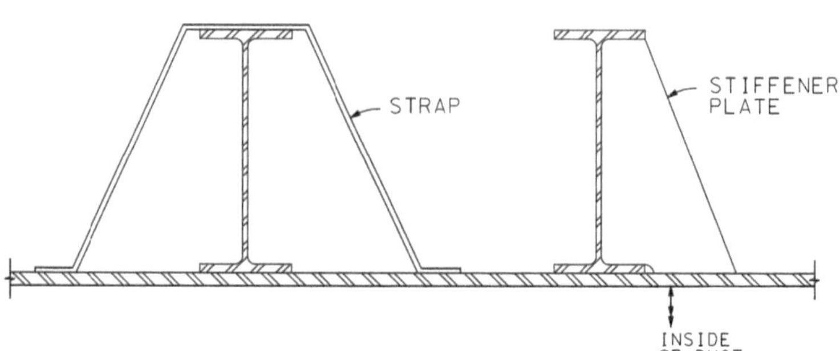

Figure 8-5. Outboard flange bracing.

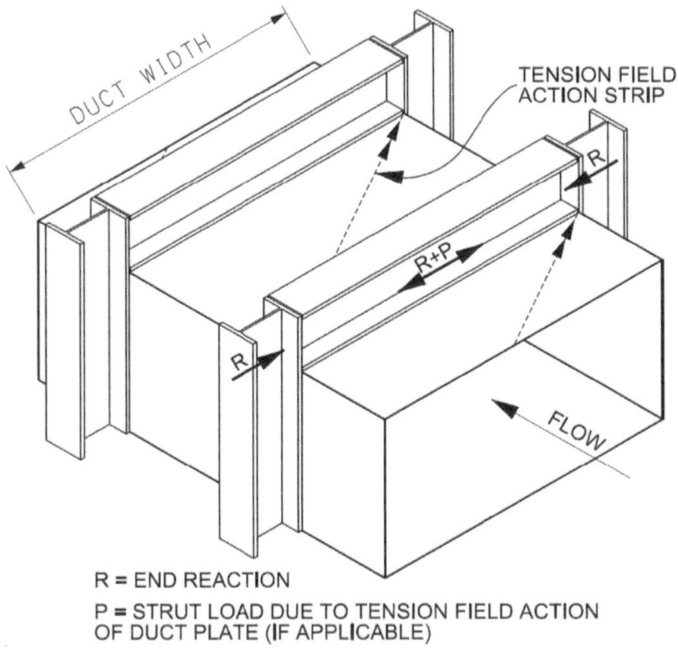

Figure 8-6. Duct stiffener axial load.

deformations of the support frame such that second-order effects exist in the members. The deflected shape for a typical braced support frame is shown in Figure 8-7. The AISC's *Specification for Structural Steel Buildings* (2017a) and *Stability Design of Steel Buildings* (2013) provide guidance on how stability provisions are applied in the analysis and the design of steel members. The methods mentioned in these publications can also be used to address stability at the duct support frames.

Stiffener Sizing at Hopper Work Points. Figure 8-8 shows a condition where a lower side panel stiffener must also help form the lateral restraint for an ash hopper sloped wall. The forces from the hopper contents will accumulate and travel up the sloping plate.

The vertical component of this force must get into the vertical duct side plate. The only way this can happen effectively is to provide a means for the force to turn the corner at the hopper work point. This is done by providing a horizontal stiffener and sometimes a reinforcing corner angle that will resist the horizontal component of the force in the sloped hopper wall. The remaining force is the vertical component, which will be resisted by the side wall panel plate.

8.2.3 Stiffener End Connections

Most end connections for structural elements are welded and must be strong enough to transfer load into the adjacent wall or stiffener. Internal bracing and trusses use gusset plates to transfer their loads to certain stiffeners. Welding is

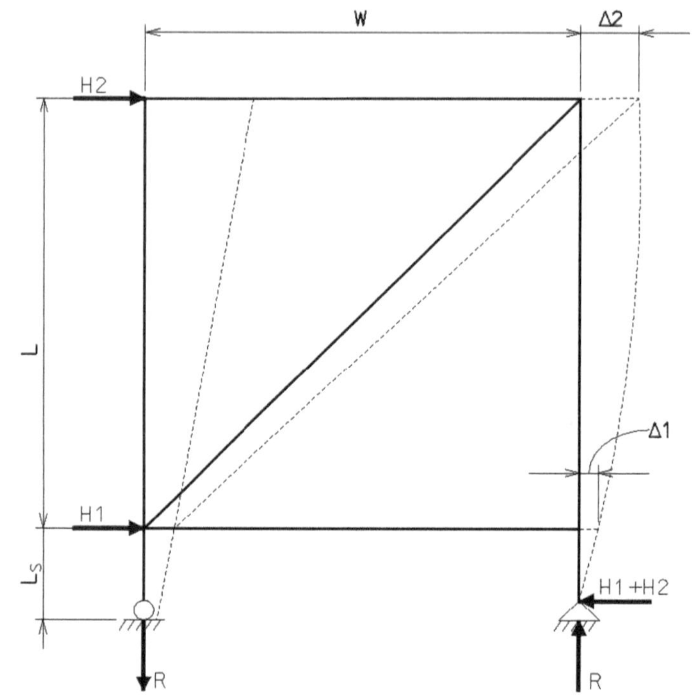

Figure 8-7. Duct support frame deformation.

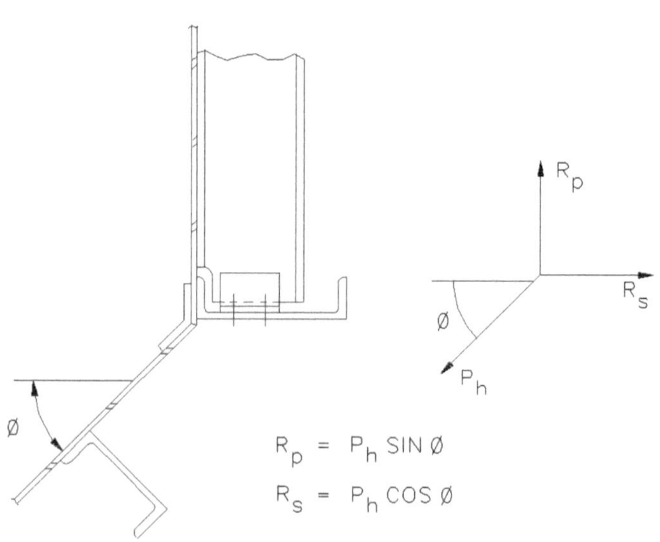

Figure 8-8. Duct-to-hopper interface.

traditionally used for most ductwork element connections. Welding gives the erector some flexibility in assembly and erection, and usually allows smaller connections. Also, in high-temperature application, welds behave better than bolts; bolts tend to loosen with continued exposure to high, fluctuating temperatures.

End connections may be designed as either pinned or fixed. In the case of single-span stiffeners, the most economical and proper design would be pinned connections, given their simplicity. This is true even though moment connections would allow the design of a lighter structural member. As in many cases, the structural engineer is usually better off with a straightforward and predictable approach to the design process, because the result is more reliable. Also, end connections designed as pinned connections will be smaller and less rigid and have less welding. Stiffer connections will tend to develop secondary stresses from strain-controlled loadings.

Pinned Connections. Treating an end connection as though it has zero moment is a simplification of its real behavior. Some simple connections may have up to 20% fixity. However, permissible local yielding and deformation will still allow the stiffener to develop its full design capacity as a simply supported member.

There are many different ways of arranging pinned connections, and Figure 8-9 shows some typical arrangements. In the design, the objective of a pinned connection is to provide an adequate path for the end reactions to be transferred into the adjacent stiffener or directly into the adjacent plate without developing appreciable end moments. Each element of the connection, including the welds, should be designed to transfer all loads without being overstressed.

Shear tab or knife plate connections should be designed to have enough ductility to be considered pinned. This is achieved by limiting the plate thickness so that the plate moment strength is less than that of the connecting element, whether it is bolts or welds. Guidance is provided in the AISC *Steel Construction Manual* (2017b) Part 10, and other structural engineering textbooks.

Moment Connections. Moment connections are used to form a continuous frame structure around the duct so that moments can be transferred from one stiffener, around the duct corner, to another stiffener. These moments help limit stiffener sizes by maintaining a negative moment balance at the end of each stiffener that counters some of the positive moment at the midspan of the stiffener. Moment connections are also used on stiffeners that serve double duty as duct supports. Sometimes, portal frames are used at duct support locations where sidesway must be resisted and where internal trusses are not desired. These duct portal frames transfer wind and other lateral loads to the duct support points. Moment frame configurations are highly indeterminate, and the relative stiffnesses between the adjacent sides must be accurately accounted for in the analysis and design. The most efficient designs provide a comparable relative stiffness between adjacent members so that the weaker member does not distort or become overloaded. Figure 8-9 shows examples of two types of moment connections.

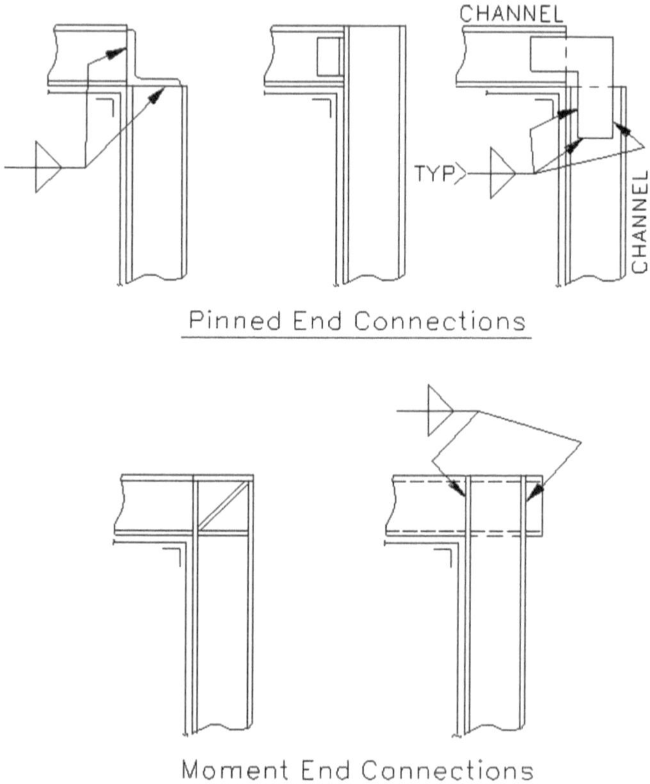

Figure 8-9. Examples of stiffener connections.

Moment connections take much more effort to design and fabricate than pinned connections. Calculating the design loads of a moment connection is typically done when the stiffeners are sized. Because the duct stiffeners and their connections act together, the analysis of the system could be complex, and the use of a computer program should be considered. An alternative would be to design moment connections that develop the entire capacity of either of its two members. However, these connections may then be considerably overdesigned and expensive.

Moment connections should consider any potential effects of thermal gradients. This is of particular importance at a moment frame, where the restraint at the corners develops stresses within the support stiffener members (Figure 8-10). The stiffer member of the frame deflects because of the thermal gradient, but the corner connections are rigid and rotation cannot occur, so this restraint causes the less stiff member to deflect opposite the thermal gradient strains, resulting in a uniform moment in the moment frame stiffeners. The magnitude of the moment depends on the moment of inertia and span of the support frame members.

Moment connections are also labor-intensive to fabricate and erect, because they may require the use of reinforcing plates and considerable welding.

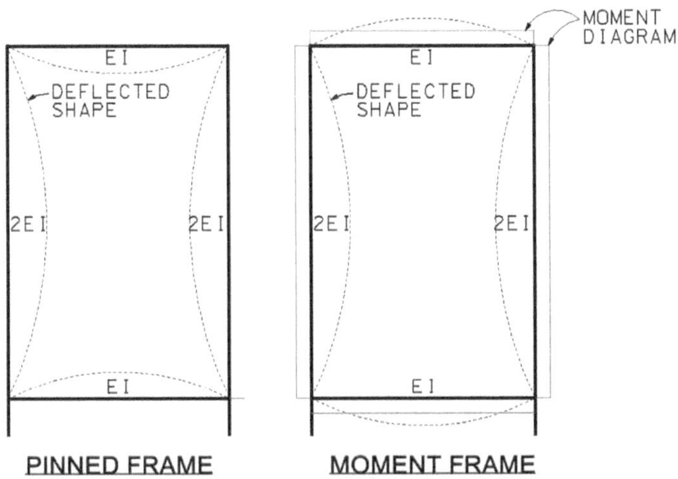

Figure 8-10. Typical thermal gradient effect on moment frame.

Reinforcing plates would be needed at locations where stiffening is required to develop the full-strength capacity of the connection. The structural engineer should be aware that even though end connections may be moment connections in the plane of the analysis, they may need to be considered as pinned connections in the out-of-plane direction.

8.2.4 Corner Configurations

The plate edge support assumptions that are used in the structural analysis must agree with the actual corner configurations used. The structural engineer must be aware of these details and must understand their structural behavior. Many types of corner configurations are used in rectangular ductwork. An angle embedded in the corner is the most common. Stiffeners may frame around the duct corners, or they may stop short at the corners. An angle or bent plate embedded in the duct corner allows erection adjustability, serves as a portion of the flange of the wall, roof, and floor panel diaphragm system, and/or serves as a support seat for the duct stiffeners. Corner angles should be designed for the corresponding axial forces when they are considered in the analytical model selected by the engineer, such as the chord members in a Pratt truss analytical model, as discussed in Chapter 6.

The design of the corner angle typically considers the composite section properties (Figure 8-11). The effective duct plate *kt* is similar to that of a duct stiffener and is discussed in greater detail in Section 8.2.2.

The corner may simply be a vertical plate panel fillet welded to a horizontal plate panel with no corner angle. Whatever the arrangement, the load path from the plate to the corner, from the stiffeners to the corner, and from stiffener to stiffener must be thought out thoroughly by the structural engineer so that the plate analysis approximates reality.

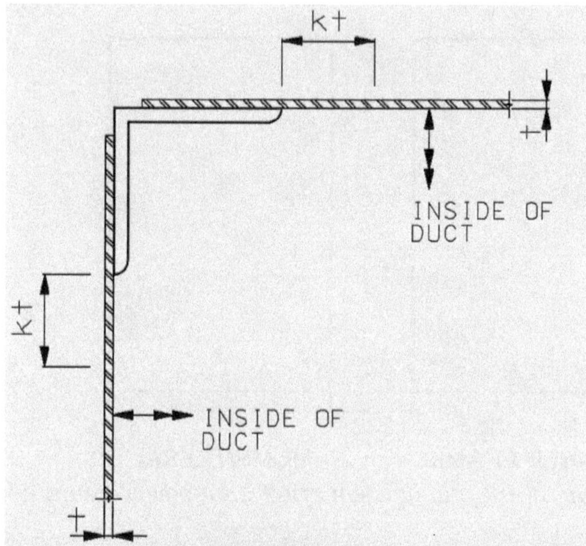

Figure 8-11. Corner angle composite section.

Generally, duct corner angles simplify field assembly, improve the duct's overall torsional strength, improve the duct's overall ability to support itself through girder or truss action, and increase the duct section stiffness. This greater stiffness is particularly effective in a dynamic environment.

8.3 CIRCULAR DUCTS

As discussed in detail in Chapter 2, circular ducts are often used in locations where there are straight duct runs with sufficient space and a conveniently located support structure. Circular ducts are designed with some of the same types of structural elements as rectangular ducts, but there are differences, which will be explained here.

8.3.1 Ring Stiffeners

Just as in the design of pressure vessels, stiffener rings for circular ducts are more efficient at retaining internal pressure force than straight stiffeners on rectangular ducts. In circular ducts, the internal pressure is resisted by axial hoop stress in the shell. Additional economy is realized by the lack of end-connection hardware, but the cost of splicing and bending the stiffener rings should also be considered.

The design of ring stiffeners is different from the design of rectangular duct stiffeners because ring stiffeners resist internal pressure loads through hoop action that results in axial forces in the ring stiffener, whereas in rectangular ducts the stiffeners resist internal pressure loads through flexure. The ring stiffener design should also consider the axial force and bending moments from the tributary wind or seismic loads. More guidance is provided in Troitsky (1991, Chapters 5 and 9).

Ring stiffeners should be welded to the duct plate so that the pressure loads can be transferred to the stiffener and so that there is enough horizontal shear transfer to form a composite section. As with rectangular ducts, axial forces will be carried by this composite section through its centroidal axis. The effective length of the duct plate may be taken as $l_s = t_w + 1.664\sqrt{Rt} + \frac{b_f t_f}{(t + t_f)}$, as defined in Chapter 9 of Troitsky and in V. M. Klykov's *Investigation of the Supporting Rings at Pipeline of Large Diameter* (1958), where t_w is the web thickness of the stiffener, b_f is the width of the stiffener flange, t_f is the width of the stiffener flange, R is the cross-section radius of the ductwork, and t is the duct plate thickness.

Ring stiffeners also serve as the attachment for the duct supports. Figures 8-13, 8-14, and 8-15 provide examples of circular duct support schemes. The composite section of the stiffener ring should have the strength to resist both the forces from internal pressure and the additional local forces and moments from the duct supports. These forces and moments can be determined using common circular ring formulas, such as those in *Roark's Formulas for Stress and Strain* (Young et al. 2011), Troitsky (1991), and the ASME *Boiler and Pressure Vessel Code* (2015a). The effects of internal pressure and support stresses can be combined by superposition. Figure 8-12 shows an example of the local moments and forces that are developed at a circular duct support point. These internal forces and moments are a function of the load to the supports, *W*, the radius of the ring, *R*, and the support geometry. A computer model of the

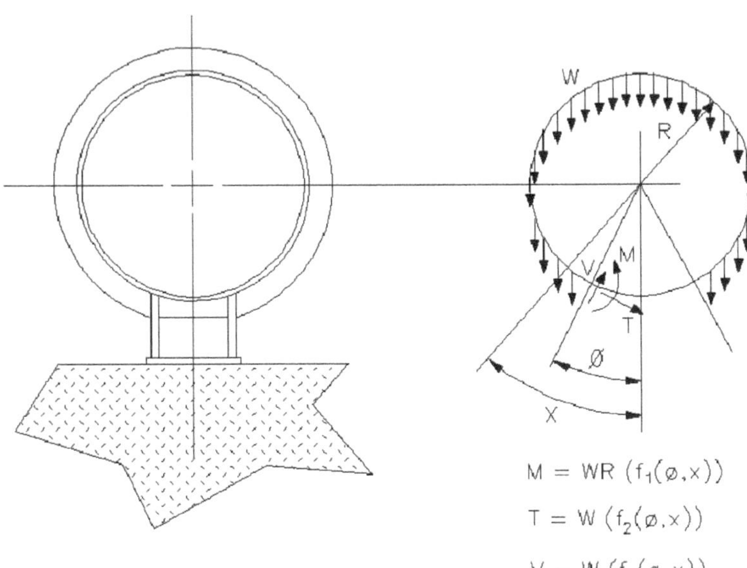

Figure 8-12. Local stresses in circular rings.

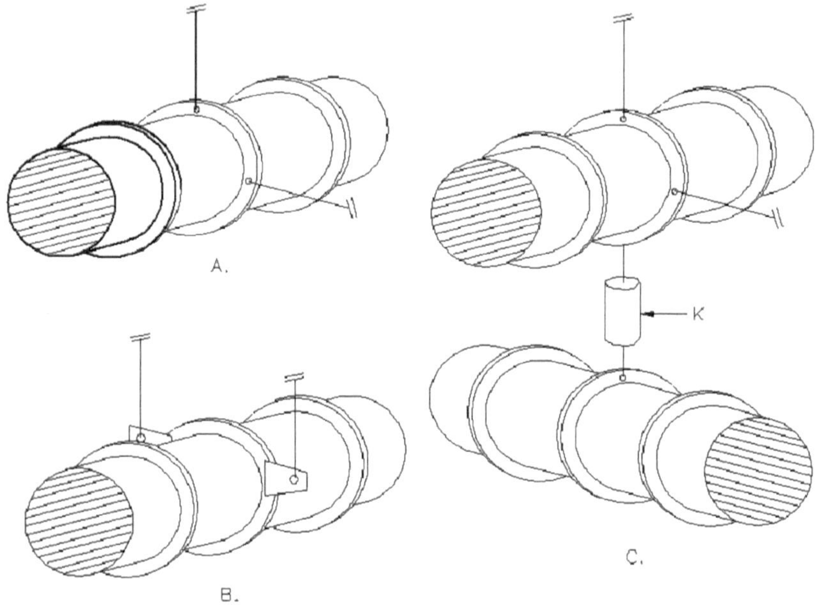

Figure 8-13. Typical circular duct support hanger arrangement.

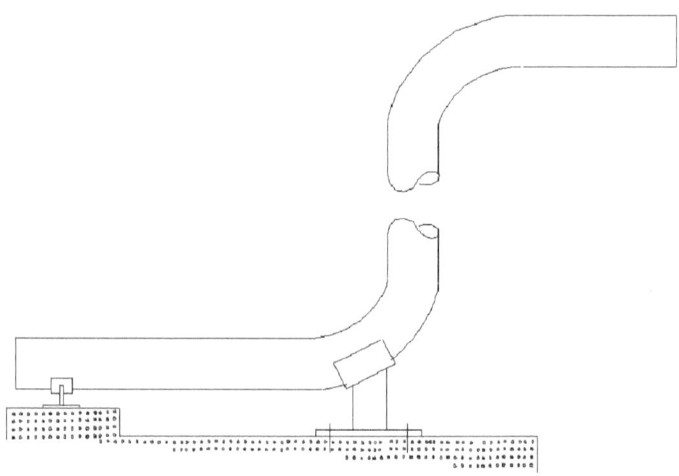

Figure 8-14. Typical bottom-supported circular duct arrangement.

duct support and the associated ring stiffener may be beneficial to determine the shears and moments in the ring stiffener.

The engineer should consider the ease with which the members can be rolled for a typical duct diameter. Typically angles, flat bars, hollow structural shapes, or tee shapes are preferred, but certain wide flange shapes can also be bent about the

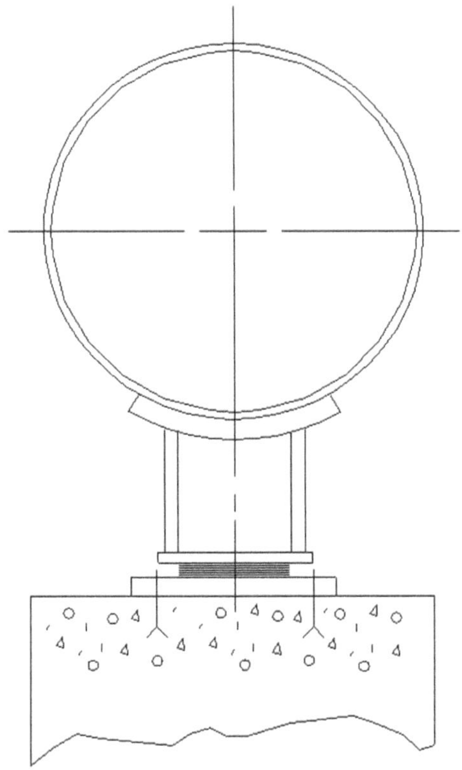

Figure 8-15. Typical circular duct saddle support.

strong axis, depending on the diameter. At certain support ring stiffeners it may be necessary to splice two tee shapes along the stem in lieu of bending a wide flange shape. Bending of channels about the strong axis is not recommended, because the asymmetrical cross section can result in distortion during the bending process.

8.3.2 Internal Struts

Internal struts in circular ducts are rare, because the inherent strength of the cylindrical section usually is more than adequate to carry the load. However, a circular duct under high external pressure may receive the most benefit from internal struts. The structural engineer should be aware that enough struts should be used to force buckling in the higher-mode shapes that correspond to critical loads greater than the applied load times a rational factor of safety. Also, sometimes internal struts are used at the support locations to reduce the plate stresses.

The decision to use internal struts versus strengthening of the stiffener section depends on various factors, including flow pressure drop requirements, clearance requirements, temperature differentials, and flow-induced vibration. The structural engineer should be able to recognize all the factors involved and coordinate efforts to arrive at a design acceptable in both structural and functional aspects. In addition to functional design conditions, internal struts can be used as temporary bracing to control ovaling during shipping and manufacturing. Temporary bracing is discussed in further detail in Chapter 10.

8.4 INTERNAL TRUSSES AND STRUTS

Lateral loads on ducts need to be transferred to the structural support points, and structural continuity is required across duct openings; internal trusses are the most direct way of achieving the proper load transfer and stiffness at the duct openings. Ductwork internal trusses and struts are also used to provide intermediate support of stiffener members. This technique allows the use of longer stiffeners with smaller cross sections. The structural engineer should ensure that the connections of truss members and intermediate struts are adequately attached to the ductwork external stiffeners, which usually serve as the truss chords, by penetrating the duct plate through a cut-out and welding directly to the stiffener. A detail of this concept is shown in Figure 8-19. Instances where internal trusses and struts might be used are discussed in Chapter 2.

A section of duct could have a complex geometry because of limited space or because it transitions into a major piece of equipment. This may occur when new equipment is designed to be installed among existing equipment. In these cases, internal trusses are helpful in transferring the loads around the corners and through the transitions.

The decision to use internal trusses or struts usually involves some trade-offs. The structural engineer faces higher costs for internal elements but lower costs for the stiffeners. The structural engineer should recognize the functional impact of using internal bracing, because the mechanical performance of the duct will be degraded by a larger pressure drop. Pressure drop calculations usually have some additional allowance for undefined internal obstructions, but the structural engineer should provide the mechanical process engineer the internal bracing locations, arrangement, and member sizes, including any associated connections, so that these obstructions can be directly input into the computational fluid dynamics or physical flow model. More pressure drop through the system requires more auxiliary power to run the unit and thus higher cost. Over the life of the plant, the additional operational cost from the increased pressure drop may exceed the material cost savings from using internal trusses. Ideally the number of transverse internal trusses should be kept to a minimum to reduce the pressure drop through the flue-gas system.

Internal trusses and struts are in the flue-gas flow, and so these elements may be corroded or eroded over the design life of the duct. An allowance on the thickness of these members may be required based on the flue-gas environment. The required allowance depends on the flue-gas flow, the fly ash particle size, the fly ash chemistry, the fly ash flux (pounds of fly ash per square foot per hour), the design life, and so on. Determining the actual degree of erosion is a complicated and inexact science. The structural engineer should rely on experience or plant data to supplement any calculations prepared for the anticipated erosion loss. A sacrificial angle, plate, or pipe cradle may be added along the leading face in lieu of using an erosion allowance. The sacrificial member should not be included when determining the internal truss strength but should be included when determining the dead load of the structure.

Single struts are often used as intermediate supports or span breakers for duct stiffeners. These struts are typically analyzed and designed to carry only the equal pressure reactions from both sides of the duct. With symmetrical stiffener framing, the pressure load cancels through the struts, and therefore the moment in the duct stiffeners is reduced. Sometimes struts are used to allow the stiffeners on both sides of a duct to share in the resistance to the applied external load, such as dead, ash, live, or wind. The impact on the duct stiffener in the presence of a strut is shown in Figure 8-16. The use of span-breaker struts can drastically reduce stiffener sizes.

8.4.1 Analysis Methods

The structural analysis of a truss can be performed using classical manual calculation procedures if the truss is statically determinate. Trusses are typically analyzed assuming that all truss elements have pinned ends. This is typically the case when gusset plates are used in the connections, as shown in Figure 8-19. If truss diagonals are connected in a manner that cannot be assumed by the structural engineer to be pinned, then the analysis should be performed considering the rotational rigidity of the connections. The moments from this analysis should then be considered in the design of these diagonals.

If the structural engineer decides that an indeterminate structural analysis is worthwhile because of member end conditions that cannot be assumed to be pinned, a computer analysis should be performed using a verified structural analysis program. In this analysis, the structural engineer must be sure to model the end connection condition that exists in the plane of the analysis, even though the connection may have a different condition in the out-of-plane direction. For most truss analysis applications, especially when there are many loading combinations to investigate, the most convenient and efficient tool is a verified structural analysis and design computer program.

8.4.2 Truss Element Design

As discussed in Section 8.2.2, ductwork support frames or trusses are similar in behavior to braced frames in a building. Forces, moments, and displacements exist within the duct support frame or truss from second-order effects to maintain equilibrium in the deformed state. The AISC specification requires the analysis to consider the influence of second-order effects (P-Δ and P-δ), deformations, geometric imperfections, and a stiffness reduction from residual stresses in the member. Any member or connection deformations that contribute to the overall

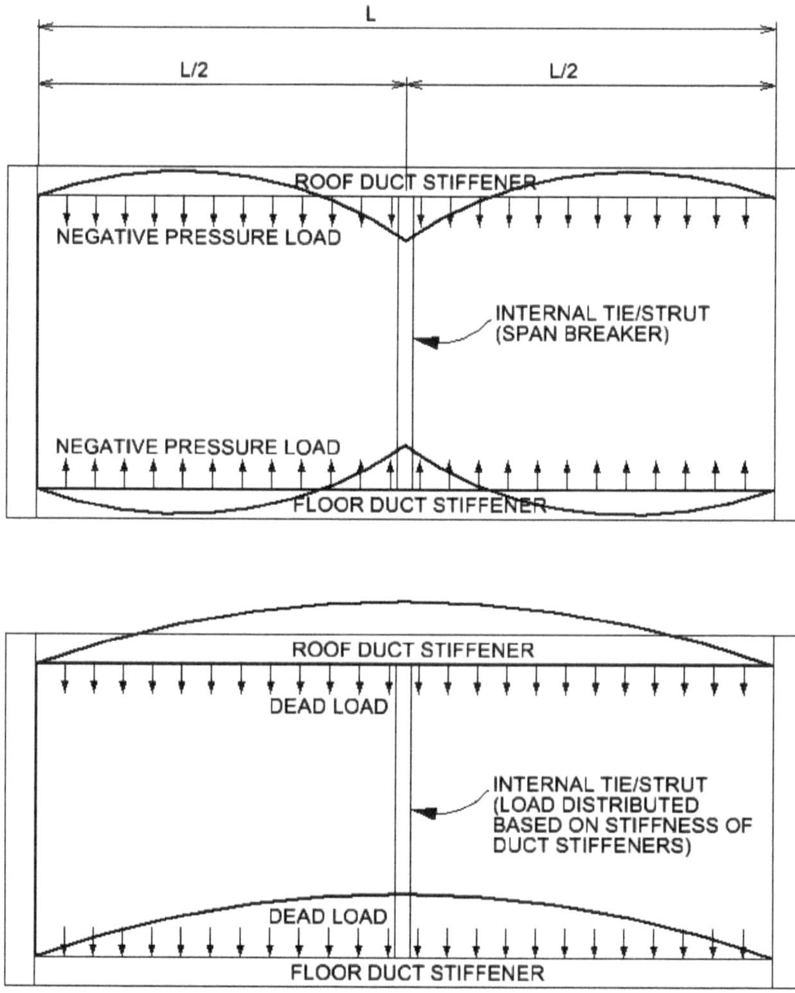

Figure 8-16. Moment diagram with a single strut.

lateral displacement of the structure should be considered in the analysis. Guidance on satisfying the stability provisions is provided by AISC Design Guide 28, *Stability Design of Steel Buildings* (2013) and AISC *Specification for Structural Steel Buildings* (2017a). These stability requirements are applicable to the entire structure, including the compression members, beams, bracing, and all associated connections.

Internal Elements. Typically, internal members have only tension or compression forces, assuming that the member has been analyzed with pinned ends and it is loaded through its neutral axis. Where full or partial moment connections are used or the member is eccentrically loaded, the worst combination of axial load and bending moment needs to be considered. As recommended in the AISC specification, it is preferable for compression elements to be restricted to a maximum KL/r ratio of 200 and tension elements, other than rods, to a maximum L/r ratio of 300.

Depending on the analysis and design methods, forces in internal members as a result of the second-order analysis and/or notional loads may need to be combined with the first-order member forces.

Duct Stiffeners. Duct stiffeners are typically used for the chord members of the internal trusses. As with other duct stiffeners, bending moment in the duct stiffener at an internal truss is caused by pressure, ash, dead, live, wind, and seismic loading. The difference is that the duct stiffener will be designed as a continuous beam between truss panel points. The duct stiffeners at an internal truss should be designed for the combined effect of the bending moments and the truss chord axial forces. The result is that there is usually more than one critical section that must be evaluated on each member.

Vibration Considerations. There may be additional constraints on the stiffness of truss elements if the elements are subject to flow-induced vibration. When an element is placed in cross flow, vortexes are shed periodically from the element at a rate called the vortex shedding frequency. If this approaches the natural frequency of the element, the two frequencies can become synchronized (locked in), and high-amplitude vibration of the element can occur. This can lead to fatigue failures.

The vortex shedding frequency can be analyzed directly if the flow velocity, density, and viscosity are known. It may be compared with the natural frequency of the member in a direction perpendicular to the flow. Multiple modes of vibration and multiple flow velocities may need to be evaluated, depending on the number of potential operating conditions.

Vibration caused by vortex shedding is discussed in various publications, such as *Harris' Shock and Vibration Handbook* (Harris and Piersol 2002) and *Flow-Induced Vibration* by R. D. Blevins (2001). The vortexes impart lift forces to the internal structural element at the vortex shedding frequency in the direction perpendicular to the flow. Drag forces are applied in the direction parallel to the flow at twice the vortex shedding frequency (Blevins 2001). Typically, the amplitudes in the perpendicular-to-flow direction are larger than those resulting from drag forces in the parallel-to-flow direction. However, the parallel-to-flow direction may need to be analyzed if the natural frequency is lower in that direction (example would be the weak direction of an I-beam).

The vortex shedding frequency is calculated as

$$f_S = \frac{\mathrm{St} \times V_g}{b}$$

where

f_s = Vortex shedding frequency of the air or gas,
b = Width of the member exposed to flow,
St = Strouhal number, and
V_g = Flow velocity of the air or gas.

The Strouhal number for various cross sections can be found in hydraulic textbooks, such as Blevins (2001), and in European Committee for Standardization (2010), Standards Institution of Israel (2008), and ASCE Task Committee on Wind Forces (1961). The Strouhal number for commonly used cross sections of truss elements in ductwork is shown in Figures 8-17 and 8-18. Note that the Strouhal number for noncircular cross sections may depend on the orientation of the cross section to the flow field (angle of incidence).

The calculation of the truss element's natural frequency should be based on simply supported end conditions, unless the evaluation of the connection details and the structural element arrangement can justify full or partial rotational end restraint. Note that the orientation of gusset plates at connections can cause the natural frequency of the truss element to vary in the perpendicular-to-flow and parallel-to-flow directions. The gusset plates can also create localized flow turbulence if they are large and oriented perpendicular to the flow. Also consider that large truss elements and connection plates can block off enough flow area to significantly increase the flow velocity across the localized area of the truss. The material properties used to determine the natural frequency should be adjusted to reflect the duct operating temperatures.

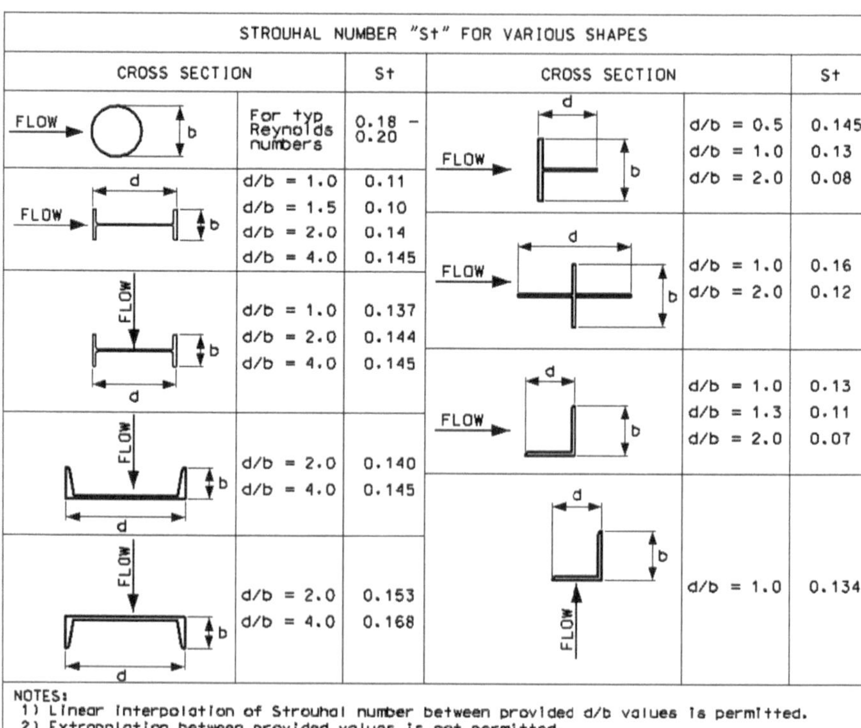

STROUHAL NUMBER "St" FOR VARIOUS SHAPES

CROSS SECTION		St	CROSS SECTION		St
Circle (FLOW, b)	For typ Reynolds numbers	0.18 – 0.20	T-section (FLOW, d, b)	d/b = 0.5	0.145
				d/b = 1.0	0.13
				d/b = 2.0	0.08
I-section, flow along d (FLOW, d, b)	d/b = 1.0	0.11	Cruciform (FLOW, d, b)	d/b = 1.0	0.16
	d/b = 1.5	0.10		d/b = 2.0	0.12
	d/b = 2.0	0.14			
	d/b = 4.0	0.145			
I-section, flow across d (FLOW, d, b)	d/b = 1.0	0.137	Angle (FLOW, d, b)	d/b = 1.0	0.13
	d/b = 2.0	0.144		d/b = 1.3	0.11
	d/b = 4.0	0.145		d/b = 2.0	0.07
Channel, flanges toward flow (FLOW, d, b)	d/b = 2.0	0.140	Angle, flow from below (FLOW, d, b)	d/b = 1.0	0.134
	d/b = 4.0	0.145			
Channel, flanges away from flow (FLOW, d, b)	d/b = 2.0	0.153			
	d/b = 4.0	0.168			

NOTES:
1) Linear interpolation of Strouhal number between provided d/b values is permitted.
2) Extrapolation between provided values is not permitted.

Figure 8-17. Strouhal numbers for various cross sections.

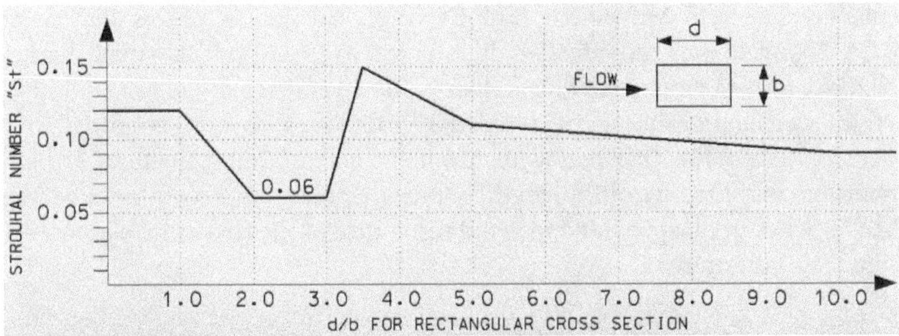

Figure 8-18. Strouhal numbers for rectangular cross section.

One effective approach is to design the truss element such that its fundamental (first mode) natural frequency is above the vortex shedding frequency. Although this is an effective design criterion, it may yield uneconomical or cumbersome designs. If designing the element to have a natural frequency above the vortex shedding frequency is not feasible, then the element should be designed with a design margin between the two frequencies. A design margin of at least 30% between the natural frequency and vortex shedding frequency is recommended by many flow-induced-vibration references to prevent lock-in.

This criterion can be used to avoid vortex shedding resonance:

$$0.7 \geq \frac{f_S}{f_N} \quad \text{or} \quad \frac{f_S}{f_N} \geq 1.3$$

where f_S is the vortex shedding frequency of the air or gas, and f_N is the natural frequency of the truss element.

Vortex shedding resonance in both the perpendicular-to-flow and parallel-to-flow directions can also be avoided or suppressed by meeting the reduced velocity and/or reduced damping values defined in ASME (2015b) Nonmandatory Appendix N article N-1300. The formation of vortexes may be minimized by streamlining the cross-sectional shape. It may also be prevented by vortex-suppression devices such as helical strakes. See Blevins for guidelines on streamlining and vortex-suppression devices. A computational fluid dynamics model is a helpful tool in assessing unique cross section and flow conditions and whether flow over the shape fosters the formation of vortexes. More complex situations may require the use of a fluid-structure interaction model and/or laboratory testing to help predict the potential for vibration of a unique system.

Sometimes restricting an internal member's span-to-depth (L/d) ratio to some maximum value will provide a section that is stiff enough that vibration will normally not be a concern. Typically, a maximum between 20 and 30 can be used. This reduces the economy of the member design, but may alleviate vibration problems.

Vibration of internal elements is especially important downstream of a fan, for a length equal to the *effective duct length*, because of the significant flow turbulence. The Air Movement and Control Association's *Fans and Systems* (AMCA 2007) defines the effective duct length as 2½ equivalent duct diameters for a 2,500 fpm (13 m/s) air flow velocity. Placing internal elements within the effective duct length should be avoided if possible, given the potential for the elements to vibrate in this region of turbulent and nonuniform flow. If placing elements in the effective duct length is unavoidable, it is recommended that a 20% to 50% margin be provided to avoid resonance of the structural elements with fan induced vibrations:

$$f_{\mathrm{N}} < 0.5f_{\mathrm{F}}\,\text{to}\,0.8f_{\mathrm{F}} \quad \text{or} \quad f_{\mathrm{N}} > 1.2f_{\mathrm{F}}\,\text{to}\,1.5f_{\mathrm{F}}$$

where f_N is the natural frequency of the stiffener or plate, and f_F is the operating frequency of the fan, or the frequency range of a variable-speed fan (note that the blade passage frequency, which is rotation speed times the number of blades, may also need to be considered).

Additional guidelines for avoiding duct pulsations and vibrations near fans can be found in AMCA (2007, Section 5.4). They include avoiding high swirling flow over vanes, dampers, or fan blades; maintaining uniform air velocity distribution; operating in the stable portion of the fan curve; ensuring adequate duct casing stiffness; avoiding standing wave patterns in the ductwork; and avoiding resonance of the duct casing with the fan running speed or blade pass frequency. Other types of flow-induced vibration, such as galloping (applicable to noncircular cross sections) and turbulence-induced vibration, may also need to be considered. Galloping can occur on flexible, lightweight, noncircular cross sections. Turbulence-induced vibration can be a concern for elements with low stiffness if the flow is turbulent. For more, see ASME (2015b) Nonmandatory Appendix N article N-1300.

Temperature Differential Considerations. If significant temperature differentials are expected within the duct, this should be accounted for by a thermal loading condition in the analysis. These thermally induced forces in truss elements should be added to the primary forces and compared to the design strength of the element.

8.4.3 Truss Member End Connections

The structural engineer should specify details that weld the gusset plate directly to the duct stiffeners. This is typically achieved by notching the duct plate around the gusset with sufficient clearance for a direct welded connection. The notch should be seal welded to the duct stiffener to prevent leakage. This detail will provide a favorable load path directly from the diagonal into the stiffeners, which usually act as the truss chords.

Most structural engineers treat truss elements as pinned members. When fixity is assumed in the analysis, the resulting moments should be considered in the connection design, and the connection detailing should ensure that adequate stiffness is provided consistent with the analysis.

The simplest diagonal connection uses a welded gusset plate for attachment of the member to the duct stiffeners. This is a low-cost way of connecting truss elements. Figure 8-19 shows a corner gusset plate to illustrate a simplified design method for a generalized loading. The welds connecting the truss diagonal to the gusset should be sized for the axial load, F_o. The welds for the gusset plate to the stiffeners should be sized for the horizontal (R_h) and vertical (R_v) components of F_o. The thickness of the gusset plate can be determined in the same way as the weld sizes. To not block too much of the flow, gusset plates should be designed with their dimensions minimized, at the cost of additional thickness if necessary. Alternative connection design methods, such as the uniform force method, may be used at the discretion of the structural engineer. Additional guidance is provided in AISC (2014) Design Guide 29, *Vertical Bracing Connections – Analysis and Design.*

There is no distinct load path within a connection. The structural engineer may choose any internal distribution of loads within the connection, as long as equilibrium is satisfied. The connection design should account for any possible limit states that may result from the chosen load distribution. Local deformations in the gusset plate and welds allow the loads to redistribute such that they closely match the assumed loading pattern.

8.5 DUCT SUPPORTS

The primary purpose of vertical duct support elements is to transfer vertical loads from the duct to the supporting structure. At the same time, these supports may also

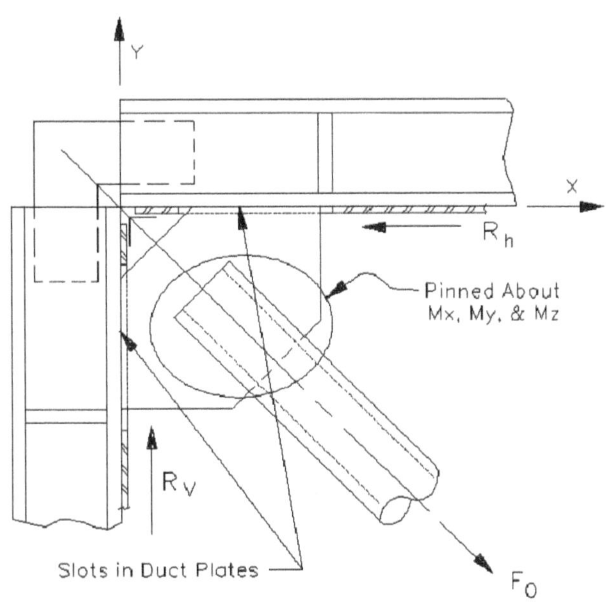

Figure 8-19. Typical truss connection.

be used to resist pressure and external horizontal loads. There are two types of vertical supports: hanger rods and bottom supports. The choice resolves down to convenience and economics. If the steel superstructure extends to elevations above the duct, then hanger rods would be a convenient method of support. However, with hangers, duct external tie elements or bumpers are required to carry the duct's lateral loads. If extra steel is needed to extend above the duct, or if the duct is within a few feet or meters of the foundation, then bottom supports may be more economical.

Hanger Rods. Hanger rods support a duct from a superstructure above it. Connections to the superstructure may be made with commercially purchased clevises or with fabricated plates. Connections to the duct may be made with simple plates welded to the duct side plates or with wing plates that are welded to stiffeners and extend out away from the duct to avoid clearance problems (Figure 8-20).

Hanger rods may be used in conjunction with spring hangers. Spring hangers can provide constant support or variable support. If spring hangers are used, they must be sized for the forces from an accurate structural analysis. If not, there could be problems with the spring hanger performance and with the duct behavior.

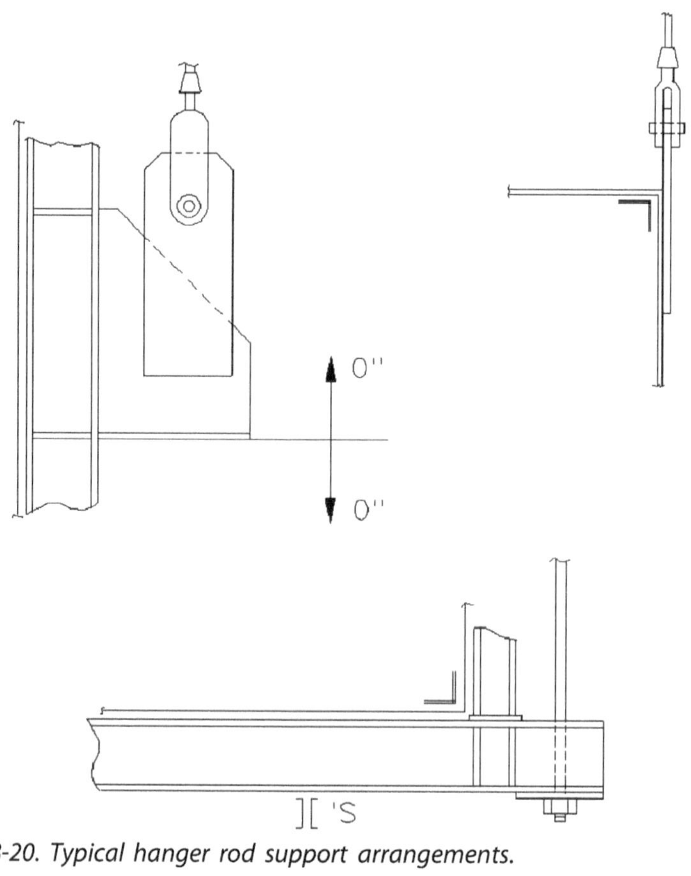

Figure 8-20. Typical hanger rod support arrangements.

When hangers are used to carry the vertical load of a duct section, external bumpers or lateral tie elements should be included to take the duct's lateral loads to the support structure.

Bottom Supports. Bottom-supported ducts use independent members or stiffener extensions to post down to the support structure. The posts should be designed for all vertical loads as well as the bending moments and shears from the lateral loads that go through the posts down to the support structure. Slide bearing plates are typically provided under the posts, except at the duct anchor point, to reduce the friction loads from thermal expansion of the duct. The posts should be designed for the moments and shears caused by these friction forces, as well as by other lateral loads, such as wind, seismic, and unbalanced pressure (Figure 8-21).

The interface between the duct support post and the support steel should account for any eccentricity from the thermal expansion of the duct. A typical practice is to offset the duct support post a distance equal to half the normal operating thermal movement. The eccentricity travels to the opposite side of the support steel centerline when the duct thermally expands. This results in an eccentricity equal to half the operating thermal movement under both normal operating and offline conditions.

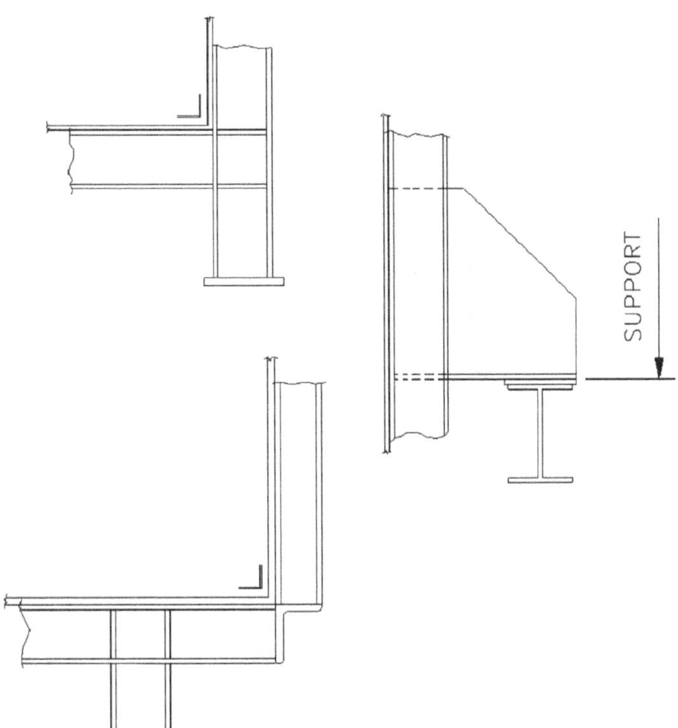

Figure 8-21. Typical bottom-support arrangements.

8.6 LATERAL EXTERNAL TIE ELEMENTS

Lateral ties are used to restrain ducts in the horizontal direction against lateral loads and to guide thermal growth. Lateral ties are external to the duct and laterally attach the duct to the support structure. Ideally, lateral ties would be located to adequately restrain a duct without having to resort to overturn couples in the vertical supports. However, this approach is usually impractical and uneconomical and therefore often is not done.

Just as with other forces, lateral loads tend to take the stiffest load path to an external restraint. Simple statics proves that the closer a load is to a restraint, the more of the load will go to that restraint. As with vertical loads, there are different methods used to distribute lateral loads to the restraint system in a rational manner. Loads that are applied overall to the duct may be spread evenly to the restraints via the duct plate and plate stiffeners. Loads that are concentrated in one area may be distributed to only the nearest restraints. Lateral ties are almost always pure tension or compression elements. They should be designed accordingly for all of the expected loads and loading combinations.

8.7 SERVICEABILITY AND DEFLECTION LIMITS

8.7.1 Duct Plate and Stiffeners Serviceability Criteria

Maximum stiffener deflection criteria, such as the typically used L/360, are generally not as critical in ductwork design, because ducts are not occupied structures where serviceability limits must be satisfied. The exception is for the duct plate and stiffeners within the effective duct length of the fan discharge or where gunite, flake glass lining, or another coating system is to be applied for corrosion and/or erosion protection. The vinyl ester or epoxy resins in a coating system may have specific deflection and rotational requirements or strain limits to ensure adequate bonding to the subsurface under all service conditions.

Deflections can also be very important and should be considered by the structural engineer in dynamic environments. Like internal truss members, the duct plate and the stiffeners within the effective duct length are exposed to significant flow-induced vibrations. The natural frequency of the duct plate and the stiffeners should be at least 20% higher or lower than the operating frequency range of the fan.

8.7.2 Internal Truss Serviceability Criteria

Internal trusses, struts, and turning vanes should be avoided within the effective duct length of the fan discharge, given the turbulent flow conditions. When internal elements are required they should have sufficient stiffness to prevent detrimental deflections caused by vibrations. Like flow-induced vibrations, the natural frequency of any internals should be at least 20% higher or lower than the operating frequency range of the fan. Typical vortex shedding frequency equations may not be applicable within the effective duct length because of the nonuniform flow profile. In the past adequate stiffness has been achieved by using internals with a maximum span-to-depth ratio (L/d) of 20 or 30.

References

Air Movement and Control Association. 2007. *Fans and systems: Publication 201-02 (R2007)*. Arlington Heights, IL: Air Movement and Control Association.

AISC. 2013. *Stability design of steel buildings: AISC design guide 28*. Chicago: AISC.

AISC. 2014. *Vertical bracing connections-analysis and design: AISC design guide 29*. Chicago: AISC.

AISC. 2017a. *Specification for structural steel buildings*. AISC 360-16. Chicago: AISC.

AISC. 2017b. *Steel construction manual*. AISC 325. Chicago: AISC.

ASCE Task Committee on Wind Forces. 1961. *Wind forces on structures: ASCE transactions 126*, 1124–1198. Reston, VA: ASCE.

ASME. 2015a. *ASME boiler and pressure vessel code, Section II: Materials*. New York: ASME.

ASME. 2015b. *ASME boiler and pressure vessel code, Section III: Rules for construction of nuclear facility components*. New York: ASME.

AWS (American Welding Society). 2015. *Structural welding code, steel*. AWS D1.1. Miami, FL: AWS.

Blevins, R. D. 2001. *Flow-induced vibration*, 2nd ed. Malabar, FL: Krieger.

Blodgett, O. W. 1976. *Design of welded structures*. Cleveland: James F. Lincoln Arc Welding Foundation.

CEN (European Committee for Standardization). 2010. *Eurocode 1: Actions on structures–Part 1-4: General actions– Wind actions*. EN 1991-1-4:2005+A1, CEN/TC 250. Brussels, Belgium: CEN.

Harris, C. M., and A. G. Piersol. 2002. *Harris' shock & vibration handbook*, 5th ed. New York: McGraw Hill.

Klykov, V. M. 1958. "Investigation of the supporting rings at pipeline of large diameter." *Construction and Architecture*, (in Russian) **8**, 1958.

Standards Institution of Israel. 2008. *Characteristic loads in structures: Wind loads*. SI 414. Tel Aviv-Yafo, Israel: Standards Institution of Israel.

Troitsky, M. S. 1991. *Tubular steel structures: Theory and design*. Washington, DC: James F. Lincoln Arc Welding Foundation.

US Steel Corporation. 1981. *Steel design manual*. Pittsburgh: US Steel Corporation.

Young, W. C., R. G. Budynas, and A. M. Sadegh. 2011. *Roark's formulas for stress and strain*. New York: McGraw Hill.

CHAPTER 9

Structural Design of Flow Distribution Devices

9.1 FUNCTION OF FLOW DISTRIBUTION DEVICES

Turning vanes, splitter plates, baffles, and perforated plates are types of flow distribution devices in the duct's flow path (Figure 9-1). Their function is to control the flow of the air or flue-gas across the duct cross section.

Turning vanes and *splitter plates* have similar functions. Turning vanes are used to change the direction of the stream, reduce flow turbulence, develop uniform distribution of flow, and minimize pressure drop. Splitter plates are linear elements used to divide and direct the stream to obtain uniform distribution of flow across the entire duct section, reduce turbulence, and minimize pressure drop. Developing uniform flow distribution and reducing flow turbulence will also minimize the accumulation of ash and sludge that settle out onto the ductwork floor. Trailing-edge liquid collectors can be provided on turning vanes and splitter plates to collect and direct liquid runoff from condensed moisture in the flue-gas stream, if necessary.

Baffles and *perforated plates* are placed perpendicular to the flow. Their function is to develop resistance to the flow. A baffle is installed across part of the duct section to direct a concentrated flow stream across the entire section of the duct. A perforated plate is installed across the entire duct section to cause the air or flue-gas to flow across the duct's full cross section. Baffles and perforated plates also absorb the energy of fast-moving streams and dissipate large swirls and eddies in the flow stream, reducing turbulence. Perforated plates are most often used at abrupt changes in duct cross-sectional area, such as in the inlet and outlet ductwork of precipitators and scrubbers. Perforated plates are not used extensively because of the large pressure losses associated with their use. Baffles are seldom used in ducts. A greater pressure drop means that larger fans will be required in the system.

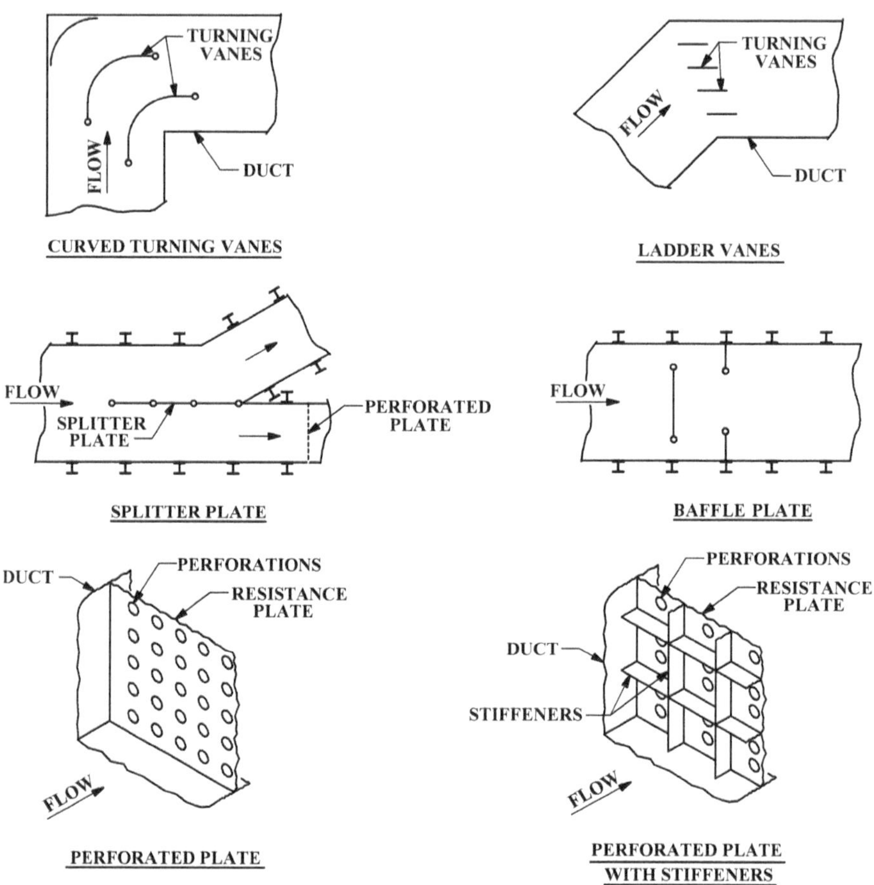

Figure 9-1. Typical flow distribution device configurations.

9.2 FLOW DEVICES FOR PROCESS EQUIPMENT

Static mixers or other flow mixing devices may be required downstream of chemical injection points. For example, static mixers are typically provided between the ammonia injection grid and the SCR box to disperse the chemical reagent across the entire duct flow cross section. Flow mixing devices can also be used to control temperature in the flue-gas system. One example is when the flue-gas temperature at the boiler economizer outlet is too low for the catalytic reaction to occur in the SCR box. An economizer bypass can be provided in this situation, and a flow mixing device is typically provided downstream of the economizer bypass duct tie-in to prevent temperature stratification across the duct cross section.

Screens may be provided to capture and remove large particles of ash. Such screens are typically provided upstream of SCRs to minimize catalyst pluggage, which may hinder SCR performance through a loss of $DeNO_x$ potential, ammonia slip, or excessive pressure drop. Ideally, ash screens are placed in sections with lower flue-gas velocity, to minimize erosion of the screen.

Internal gutters, downspouts, and other liquid flow collection devices are sometimes used in ducts where the flue-gas is below its acid dew point and excessive internal liquid condensation is expected. When this is the case, internal drains should be installed in the ducts that then carry the flow to a proper repository. Because of the corrosive nature of the liquid, these liquid flow collection devices and drains should be made from a corrosion-resistant material.

9.3 FLOW LAYOUT AND STRUCTURAL CONSIDERATIONS

The general configuration, quantity, and arrangement of splitter plates, turning vanes, and perforated plates are determined in accordance with criteria established by the system's flow analysis. The conceptual configuration and arrangement of flow distribution devices is often based on flow model testing. The testing may be done by computer simulation and/or by reduced-scale physical flow models. The structural configuration of each flow distribution device is governed by the optimum combination of the following parameters.

9.3.1 Flow Resistance

Flow resistance from splitter plates and turning vanes should be minimized. Their structural elements usually consist of plate sections stiffened with pipe sections, angles, or plates. The total projected surface of the stiffened flow device perpendicular to the flow should not significantly reduce the cross section of the duct available for flow. Limiting the total projected surface of the stiffener system to less than 5% of the total flow path is considered a reasonable control on flow resistance.

Perforated plates normally have closely spaced openings, with a ratio of open area to total area of between 30% and 75%. The opening ratio is determined by a flow analysis, which may or may not include a flow model test. Stiffeners for perforated plates may be located on the downstream surface to minimize ash or sludge accumulation and to not disrupt the flow pattern, unless the flow analysis recommends an "egg crate" arrangement on the upstream side of the perforated plate. Baffles may be perforated but are usually solid plate.

9.3.2 Physical Arrangement Considerations

The physical arrangement of splitter plates and turning vanes is usually developed to prevent intersection or interference with duct internal trusses or struts. The structural engineer must come to an agreement with the mechanical process engineer if one of these two must be relocated. If the flow distribution device must intersect a duct internal brace or truss, details should be provided to allow the independent movement of the two different structural systems. The plate elements of splitter plates and turning vanes should not be connected to intersecting members of duct internal trusses or struts unless the trusses have been designed for the loads that will be transferred from the flow distribution devices. The arrangement of baffles and perforated plates should also consider their location upstream of duct internal bracing systems. If they are placed just upstream of internal bracing members, the higher flow velocity and particle concentration caused by these devices can erode the bracing members.

Flow distribution devices subjected to corrosive or erosive flow conditions are usually provided with sacrificial shielding or corrosion-erosion thickness allowances for the structural material. Increased material thickness or supplemental members added as shielding should not be considered in the evaluation of the structural capacity of the flow distribution device. Historically, an erosion and corrosion allowance of 1/16 in. (1 to 2 mm) has often been provided in the design of flow distribution devices. Flow distribution devices in wet scrubber outlet ducts and in the scrubber inlet duct sections that are exposed to backwash should be made from the same corrosion-resistant material used to cover the inside of the duct.

Placing turning vanes at duct corners should be considered so as not to cause a dead air space. (See Figure 9-1 for a curved turning vane configuration.) The dead air space can cause thermal gradient in duct plate during normal operating conditions. To avoid this, a small gap should be allowed between turning vane and duct wall. Another solution is to design the duct corner with a radius. See Section 2.7.6 for further discussion of dead zones causing thermal gradient stresses.

9.4 SUPPORT CONSIDERATIONS

The supports for turning vanes, splitter plates, baffles, and perforated plates must be capable of transmitting all applied forces into the duct main structural system and preferably should be detailed for ease of assembly inside the duct section. Turning vanes, splitter plates, and baffles are usually attached to the duct plate. Flow distribution devices that are not used as an integral part of the duct's global structural system need only be connected to the duct plate. The structural engineer should develop connection details that distribute their reactions to the duct plate so that stress concentrations at the attachment location are minimized. Turning vanes, splitter plates, and baffles that are integral parts of the duct's structural system are usually attached directly to duct stiffeners, bypassing the duct plate. The structural engineer should develop details that transmit their reactions directly to the duct stiffener system, avoiding through-thickness loading of the duct plate. If this is not done and the flow distribution device is directly attached to the duct plate, localized bending of the duct plate should be checked.

Perforated plates are usually suspended within the duct cross section to preclude interaction of global duct forces with the diaphragm formed by the perforated plate. Isolation of the perforated plate element from the duct structure is also necessary for the efficient operation of rapper elements that are used to prevent particulate buildup. Connection details must be able to transmit the static and dynamic reactions of the perforated plate to the duct structure without significant restraint on the duct structure and without loading the perforated plate structure. Baffles are also sometimes installed in this manner.

9.5 STRUCTURAL ANALYSIS

The structural analysis and design of turning vanes, splitter plates, baffles, and perforated plates needs to address stiffness requirements, dynamic loads, static loads, and minimum loads due to local effects. These requirements are discussed in the following subsections.

9.5.1 Stiffness Requirements

Rolled structural shapes, pipe sections, and plates are the structural elements of flow distribution devices. Flow distribution structural elements in the flow stream are subjected to two possible sources of vibration excitation: *fan-induced* vibrations and *flow-induced* vibrations.

Fan-induced vibrations are a concern immediately upstream and downstream of the ID, FD, and PA fans or other large fans, as discussed in the following. The pressure downstream of the fan is postulated to contain a harmonic component at a multiple of the fan's operating speed. Flow-induced vibration is caused by the generation and collapse of eddy currents in the flow around and over the internal structural elements. The structural engineer may decide to consider flow-induced vibrations based on the service history of similar installations and anticipated flow conditions in the duct. When variable-speed fans are used, a range of operating speeds needs to be considered to establish a range of frequencies that need to be checked for resonance with the fan-induced vibrations. Fan-induced vibrations should be addressed at all flow distribution devices within the effective duct length, as defined by the Air Movement and Control Association's *Fans and Systems* (2007) or the fan manufacturer, for all large fans. The structural engineer may decide that other devices also need to be addressed.

To reduce the potential for vibrations, the use of flow distribution devices should be avoided immediately upstream and downstream of large fans. The exact requirements should be requested from the fan manufacturer. In the absence of specific criteria from the manufacturer, the design team should consider providing a constant duct section at the fan outlet, with no flow distribution devices for a length equal to the effective duct length as defined in Section 8 of Air Movement and Control Association (2007). This length is equal to 2- 1/2 duct equivalent diameters for a flow of 2,500 cfm (70 m^3/min) or less and adds 1 duct equivalent diameter for each additional 1,000 cfm (30 m^3/min) of flow. For axial fans, the length is 50% of that required for centrifugal fans, per the same publication. The equivalent diameter may be taken as 1.13 times the square root of a rectangular duct's area. Flow distribution devices should be avoided in this section of duct. Failure to meet these or similar requirements could hinder the performance of the fan or be cause for voiding of its warranty. Structural damage to the fan, the flow distribution devices, or the duct from flow-related vibrations may also result.

The natural frequency of unstiffened sections of plate elements or the leading-edge stiffener of stiffened sections can be determined and evaluated against critical operating ranges for the two possible sources of vibration. Elements susceptible to corrosion or erosion should consider all possible conditions of the element between the full and reduced cross section when calculating the natural frequency. The loss of material to corrosion or erosion may alter the element's natural frequency such that it is close to resonance with the fan or flow. The calculation of the natural frequencies is usually based on simply supported end conditions, unless the evaluation of the connection details and the structural element arrangement can justify full or partial rotational restraint. As with all duct design procedures, the material properties used to determine the natural frequencies should be adjusted to reflect the duct's operating temperatures. Stiffness evaluation methods based on deflection limits or span ratios of plates and stiffeners may also be used.

The natural frequencies of all structural elements, including flow distribution devices, should be sufficiently lower or higher than the operating frequency of the fan in the duct system so that the structural elements will not be in resonance with the fan-induced vibrations. One accepted guideline is providing a design margin of at least 20% between the natural and operating frequency of the fan:

$$f_N \leq 0.8 f_F \quad \text{or} \quad f_N \geq 1.2 f_F \tag{9-1}$$

where f_N is the natural frequency of the stiffener or plate, and f_F is the operating frequency of the fan or the frequency range of a variable-speed fan.

Flow velocities at flow distribution devices should also be significantly higher or lower than the flow velocity that causes resonance from flow-induced vibrations of the structural component of the flow distribution device. One accepted guideline is providing a design margin of at least 30% between the natural and vortex shedding frequency:

$$f_N \leq 0.7 f_S \quad \text{or} \quad f_N \geq 1.3 f_S \tag{9-2}$$

where f_S is the vortex shedding frequency (see Chapter 8).

Typically, the flow device should be designed to provide a natural frequency above the operating frequency of the fan (for fan-induced vibrations) or the vortex shedding frequency (for flow-induced vibrations).

9.5.2 Dynamic Loads

The dynamic pressure, or velocity pressure, should be considered in the design of turning vanes, splitter plates, baffles, and perforated plates. The equations used to determine the dynamic pressure are based on classical compressible flow theory.

Turning Vanes. The dynamic thrust on a turning vane or a solid baffle is

$$F_{OD} = (2\rho Q^2/A)\sin(\alpha/2) \tag{9-3}$$

The corresponding dynamic pressure is

$$SL_{DP} = F_{OD}/A_P \tag{9-4}$$

The parameters in these equations are defined in Figure 9-2 where

F_{OD} = Dynamic thrust on the duct wall or turning vane (lbs),

SL_{DP} = Dynamic pressure on the duct or turning vane (lbs/ft^2),

ρ = Mass density of the flow, usually taken as 0.002 lb-s^2/ft^4 (0.1 kg-s^2/m^4),

Q = Flow rate between the turning vanes or duct plates (ft^3/s),

A = Cross-sectional area of the duct at the bend or the area between turning vanes (ft^2),

A_P = Area of the projected duct plate or vane surface in the direction of flow (ft^2), and

α = Change in direction of the duct or vane (degrees).

Perforated Plates. One method that may be used to determine the pressure acting on a perforated plate is as follows. This equation, derived from fluid mechanics principles, is presented in Hunter Rouse's textbook, *Elementary Mechanics of Fluids* (Rouse 2011).

$$SL_{DP} = \frac{1}{2}\rho V^2/(kR^2) \tag{9-5}$$

where

SL_{DP} = Differential pressure on the perforated plate (lb/ft^2),

ρ = Mass density of the flow, usually taken as 0.002 lb-s/ft^4 (0.1 kg-s^2/m^4),

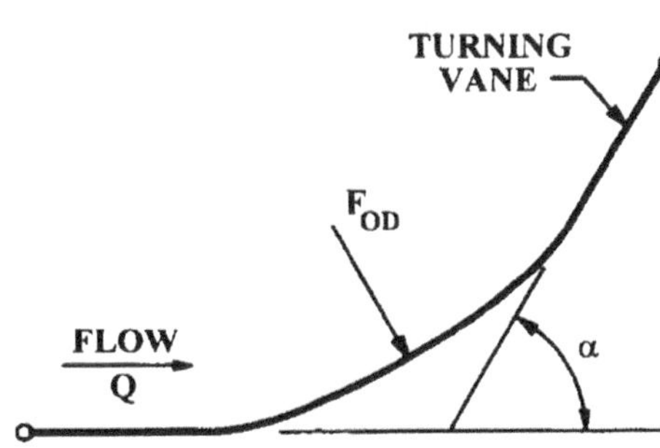

Figure 9-2. Dynamic thrust on turning vanes.

V = Velocity of the flow (ft/s),

R = Ratio of the open area of the perforated plate to the cross-sectional area of duct, and

k = A coefficient of discharge, which varies logarithmically from 0.6 at $R = 0$ to 0.75 at $R = 1$.

The differential pressure always acts on the upstream face of the perforated plate and in the direction of the flow.

9.5.3 Static Loads

Splitter plates, baffles, turning vanes, and perforated plates should be designed for all of the applicable loads used to design the main duct section. Depending on the flow velocity, ash and sludge may accumulate on flow distribution devices. Where applicable, ash or sludge loads should be applied to the horizontal surfaces of splitter plates and turning vanes, as discussed in Chapter 4. The clear distance between turning vanes or splitter plates should be used to establish the ash or sludge loading, where the load is assumed to be a function of the duct height.

Where splitter plates, baffles, or turning vanes are used as an integral member of the duct structure to resist or transmit dead, live, operational, system-related, and environmental loads, the corresponding global static loads, obtained from the wall analysis discussed in Chapter 6, are additive to the static loads of ash or sludge accumulation and the dynamic pressure forces.

The structural elements of all flow distribution devices should always be checked for compressive forces that may be induced at connections to the main duct structure when the duct is subjected to negative pressures.

9.5.4 Minimum Loads

After the selection of the initial sizes, based on stiffness criteria and dynamic loads, the structural elements of flow distribution devices may be checked for a minimum-load case. The structural elements include all plate sections and stiffeners. The minimum load is independent of the loads obtained from the dynamic pressure evaluation and the static loads. Typically used minimum values are a uniform pressure load equivalent to 25% of the duct operating pressure or 6 in. (150 mm) of water pressure, whichever is greater, to account for localized imbalances in the ductwork system. The minimum uniform pressure load is applied normal to the plane of the structural element. Reactions from this minimum load on the structural elements of the flow distribution device do not need to be considered in the design and evaluation of the overall duct structure.

9.6 STRUCTURAL DESIGN

The design of splitter plates, baffles, turning vanes, and perforated plates must consider the combination of the individual static and dynamic loads, as defined above. If a group of vanes are structurally connected, they should be treated as one structural element. As with all ductwork structural elements, the strength must be adjusted to reflect the operating and transient temperatures of the duct.

Plate sections are to be sized on the basis of forces and moments obtained from the analysis of the flow distribution device's structural systems. Plate span lengths are usually taken as the clear distance between stiffeners for flat plates, or the chord distance for curved plates. The device's stiffeners should be sized to support all contributory loads on the basis of forces and moments obtained from the analysis of simply supported structural elements, unless fixity can be justified by frame analysis.

After initial sizing of plate and stiffener sections, the dynamic response of the system to flow-induced and fan-induced vibrations is usually evaluated. Structural element size should be increased if necessary to obtain a system of sufficient stiffness to prevent resonance. When the size of leading-edge stiffeners significantly reduces the cross section available for gas flow, intermediate support plates or struts should be considered to reduce the span of the structural elements.

Where the splitter plate, baffle, or turning vanes are part of the global structural system, stiffeners on the turning vanes or splitter plates should be connected directly to stiffeners on the main duct structure so as to properly transfer the forces. Independent splitter plate or turning vane assemblies may be attached to the duct plate if their connections do not induce stress concentrations in the plate.

Special attention should be given to stitch welding near the flow device support connections. Stitch welding of vane plate to supports can result in fatigue cracks forming at these points, which can ultimately result in failure of the flow device. This is especially of concern near fans. Continuous welds should be used, as much as possible, for flow devices subjected to fan-induced vibrations.

9.6.1 Local Design Requirements

The plate and stiffener system should be sized for the combination of ash or sludge accumulation, operating pressure and dynamic pressure forces, and stiffness requirements, as defined in Section 9.4.

9.6.2 Global Design Requirements

The structural load-carrying capability of splitter plates and turning vanes may reduce the span of the exterior duct walls. In new duct design, the duct wall span is not usually reduced because of the presence of a flow distribution device. However, in some instances flow distribution devices are designed as primary structural elements that contribute to the overall duct strength. If rigidly attached to the duct walls, the structural components of the flow distribution device should always be sized for the reactions from the wall system, whether the flow distribution devices are classified as primary structural elements or not, because in actuality they will see the loads. During modification or reevaluation of duct systems, reducing the wall span based on the presence of flow distribution devices may be an effective alternative to duct stiffening. In all cases, the structural elements of flow distribution devices should be designed for any and all contributory forces resulting from the duct structural system. If possible, the typical design is to provide connections to duct wall or stiffeners that will not transfer load to the flow device. This design will also alleviate thermal stresses in the flow device that can be caused by a temperature differential between the device and duct stiffeners.

If flow distribution devices are designed as an integral part of the ductwork structure, the customer or owner should be properly informed. At a minimum, a note should be placed on the design drawings indicating this design consideration. This is done so that the owner or operator will know that if the flow distribution devices are damaged, they must be replaced, or the ductwork structure will be in jeopardy. It will also inform the owner that the flow distribution devices cannot be removed without revising the ductwork structure.

References

Air Movement and Control Association. 2007. *Fans and systems: Publication 201-02 (R2007)*. Arlington Heights, IL: Air Movement and Control Association.

Rouse, H. 2011. *Elementary mechanics of fluids*. New York: Dover.

CHAPTER 10

Drawing, Fabrication, and Construction: Techniques and Considerations

10.1 GENERAL CONSIDERATIONS

Before starting the structural analysis and design of a ductwork system, consideration should be given to the methods of shop and field fabrication, shipping, and erection. No ductwork design can be considered optimized unless these operations are given due consideration in the design process.

10.1.1 Shipping Modes and Dimension Considerations

Shipping costs are an important consideration in any ductwork project. These costs will depend on the project location, the shipping distance, and the size of the pieces being shipped. If the site allows modularization of the ductwork and shifts work that would normally be done in the field to a fabrication shop, this can save significant overall project cost and result in a higher-quality finished product.

The ductwork structural engineer should be cognizant of the shipping method to be used on the project. Because of the significance of freight cost and the importance of including this cost in the project estimate, the shipping method is often chosen prior to the engineer's involvement in the project. The structural engineer must then execute a design that can be fabricated and erected consistent with the chosen shipping method.

There are three basic modes of shipping ductwork components: rail, truck, and barge. Trucking is the most commonly used, by far, but barge and rail are sometimes practical. Depending on the project site location, one of these methods may even be necessary. The requirements of barge and rail shipping are highly specialized and should be investigated thoroughly on a project basis. Discussions in this chapter are centered on truck shipping, with some information on rail and barge.

For trucking of duct components, piece width is the critical consideration. It may be the major factor in the shipping cost. Sometimes, height and length are also significant. Weight, although important, is not usually a controlling factor. The following provides some specific information relating to cost and width.

Truck Shipment. Shipping via truck is regulated at the state level, by the individual state's governing entity, in conformance with federal guidelines. The *Permit Manual*, a condensed document published and updated quarterly by the Specialized Carriers and Rigging Association, provides invaluable information, including legal load limits and special permit requirements for oversize and overweight loads for each state, as well as contacts, legal limits, special permit limits, general restrictions, types of permits available, fees, escort needs, fines, and restricted travel areas.

As mentioned, shipping cost is a significant portion of the overall ductwork cost, and the width is usually the critical factor in the shipping cost. In general, components up to 8 ft (2.5 m) wide require no special shipping considerations; permits are required for widths from 8 ft to 12 ft (2.5 m to 3.6 m); and widths greater than 12 ft (3.6 m) require both permits and escorts. Height and length may on occasion govern the necessity for permits or escorts. Height is generally limited to 14 ft to 15 ft (4.2 m to 5.3 m) above the road surface. The maximum length, including the truck tractor and load, is generally 75 ft (23 m), but lengths up to 150 ft (45 m) are possible with permits and special shipping equipment. For comparison purposes, Table 10-1 provides rule-of-thumb cost information relative to truck shipping width, assuming no over-length or over-height conditions. This comparison was provided by Structural Steel Services Trucking in Meridian, Mississippi, based on historical data for a 40,000 lb (180 kN) load of ductwork panels shipped from Meridian to an unspecified destination 1,000 miles (1,600 km) away. The values in this table may not apply to all locations, because shipping costs vary according to the requirements of individual states. Figure 10-1 provides critical dimensions for some commonly used shipping trailers. These data may help the ductwork design team determine field erection section sizes based on permissible truck shipping lengths and heights.

Rail Shipment. Figure 10-2 shows an example of clearance requirements specified by one rail carrier. For exact limitations on width, length, and height, the rail carrier contracted to ship the material should be contacted.

Barge Shipment. Barges can carry very large ductwork sections. But such large sections may require special lifting and handling equipment at both the project site and the fabrication shop. Site lay-down area must be adequate. An experienced marine engineer should be consulted for the appropriate layout of the

Table 10-1. Cost Factors for Transport by Truck.

Shipping width	*Cost factor*
8 ft or less (2.5 m)	1.0
8 ft to 12 ft (2.5 m to 3.6 m)	1.2
12 ft to 14 ft (3.6 m to 4.2 m)	2.5

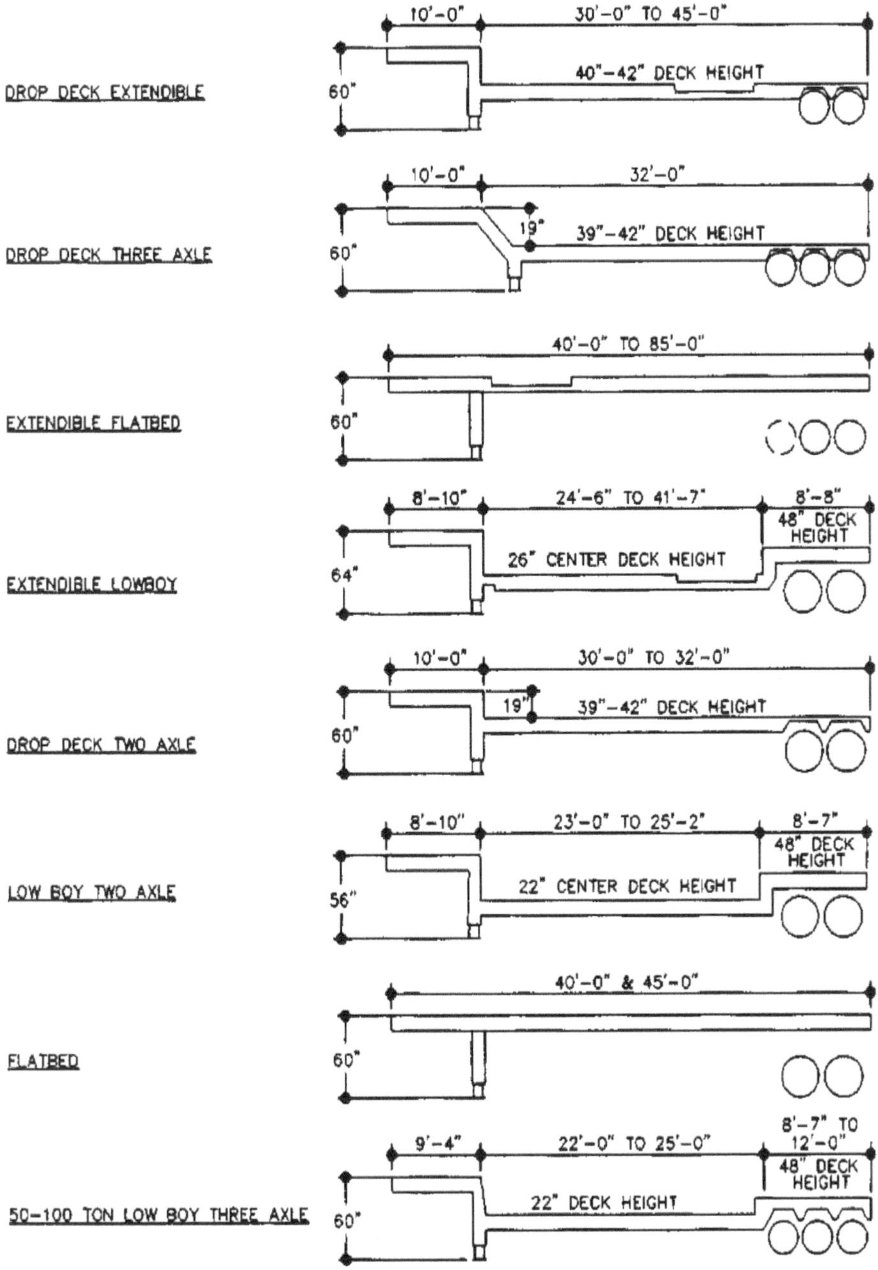

Figure 10-1. Common shipping trailer dimensions.

ductwork sections, stability analysis, and imposed loads for barge shipping. Loads on the ductwork sections will vary significantly depending on whether the barge is traveling on an inland waterway, the Great Lakes, the Gulf of Mexico, or the ocean.

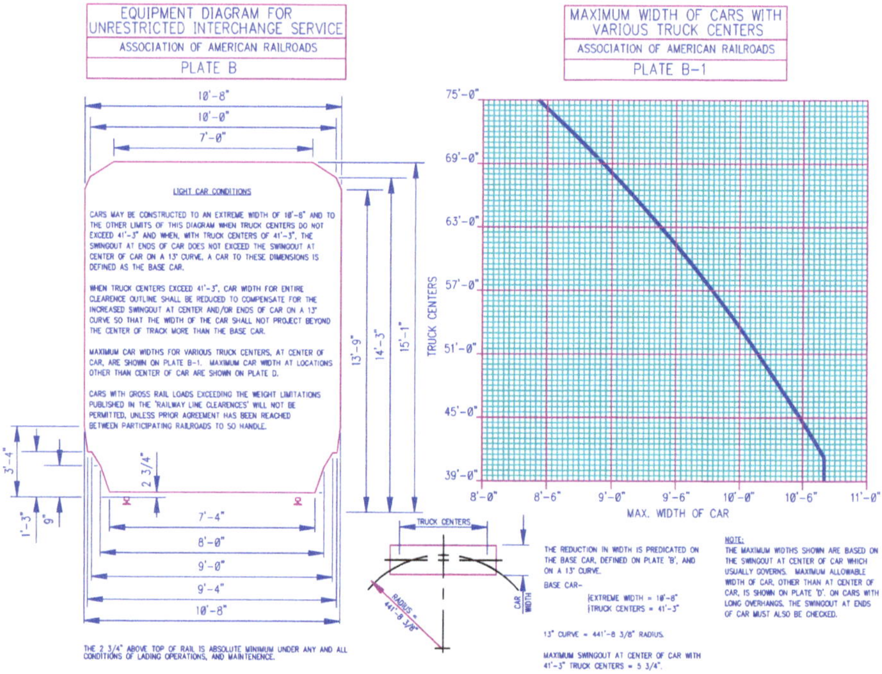

Figure 10-2. Typical rail shipping widths.

Having established the shipping method and therefore knowing the field erection dimension constraints, the structural engineer or fabricator can proceed with field splice location and stiffener placement. These field splices should be properly located and denoted on the design or fabrication drawings. See Section 10.1.3 for a discussion of field splices.

10.1.2 Construction Economy

In addition to the shipping considerations discussed in Sections 10.1.1 and 10.7.4, several other factors are important in arriving at an economical duct design: stiffener shape selection, material grade (Chapter 3), field construction costs, shop preassembly, and project site conditions.

Stiffener Shape Selection. Commonly used stiffener shape profiles are WT-shapes, channels, angles, and wide-flange or W-shapes. WTs are derived by splitting a W-shape. This splitting operation will add cost to the fabrication, but maybe not to the overall cost because of material savings.

Circular ductwork is most commonly stiffened by rolled angles, wide flanges, WTs, or bars. Most fabrication shops are equipped to roll angle shapes, but they may have difficulty rolling other, heavier shapes. Rolling shapes to small diameters may warp or crack the member or set up undesirable stresses.

The cost of rolled steel shapes varies. Prices are subject to change and should be periodically investigated by the structural engineer. The structural engineer should be aware that the profiles selected can and do influence the erected cost of the ductwork being designed. Strength and deflection considerations are not the only criteria for ensuring that the most economical stiffener is selected.

Field Construction Costs. Field construction costs are an important consideration. Unfortunately, it is sometimes difficult during the design stage to obtain comprehensive data relating to site-specific field construction costs, especially crane capacity. In many projects, the erection contractor has not been selected at the time the design is started. Thus, the data needed to make the proper determination between shop and field fabrication operations may not be available. When available, these data should be considered during the design process.

However, historically it has been understood by structural engineers that shop fabrication is less expensive than field construction. In addition to lower costs, operations such as welding, cutting, fitting, and handling are inherently of a higher quality when performed in a fabrication shop as opposed to at the project site. Therefore, it is best to specify as much shop work as possible.

Shop Preassembly. Sometimes the physical dimensions of a duct cross section permit the duct to be partially preassembled into box sections or smaller C-shaped modules in the fabrication shop. Conventionally constructed stiffened panels can be preassembled in the fabrication shop to form ductwork sections in various lengths. Shop preassembly of ductwork obviously requires more shop labor, and shipping costs are higher with these larger pieces. However, field labor, which is usually more expensive and of lesser quality, is reduced. To further minimize the field labor, insulation and lagging can also be installed in the shop before the preassembled sections are shipped to the field. Extensive shop preassembly of ductwork might also reduce critical construction time. Fabricating in the more controlled shop environment also improves site worker safety by reducing the amount of work performed above grade.

When the duct size permits one-piece shop construction of rectangular and circular ducts, the ductwork structural engineer, working with the project leaders, can determine whether box construction or panel construction is best. Considerations include shop labor cost versus field labor cost, shipping costs for panel construction versus box construction, and potential shortening of the project schedule.

If shop preassembly is done, the engineering responsibility for the preassembled pieces should be defined at the beginning of the project. When modularizing sections of ductwork, the loads imposed on the ductwork during lifting, shipping, and erection need to be considered by either the rigging engineer or the structural engineer.

Project Site Conditions. Project site conditions are also an important consideration. Any possible lay-down space constraints should be investigated. Lay-down area and site preassembly area can be scarce and should be considered in the design. Site equipment capacity and availability is also an important consideration when determining duct component shipping sizes and weight to be delivered and erected at the job site. The location of existing equipment will also play a role in determining the size of shipping pieces. Typically, the more times a piece of material is moved or handled, the more the overall cost increases.

10.1.3 Field Splices

Section 10.1.1 discussed the importance of freight cost and the structural engineer's role in defining ductwork component pieces, particularly their width and height, consistent with the shipping mode chosen for the project. Field splices, also referred to as field joints, commonly occur at ductwork corners and at panel-to-panel joints on rectangular ducts. For circular ducts, field splices occur where duct lengths are joined. Several configurations and arrangements of field splices are used. Not all configurations will be covered here. Commonly, configurations of field joints are dictated by the fabrication and construction techniques to be used on the project. Of course, design strength requirements must also be satisfied. For ductwork with insulation on the outside, it is preferred that the seal weld be on the inside. This allows inspection and repair without removing the insulation and lagging. It also puts the weld in the shear plane of the duct wall and eliminates crevices where corrosive ash could accumulate.

The structural engineer, working with the fabricator and erector, should agree on the field splice configuration best suited for the project. A common method is for the structural engineer to locate and denote the field splices on the design drawings, thus defining the sizes of the pieces to be shipped. Often the designation "F.J.," for field joint, or "F.S.," for field splice, is used on the drawings to denote the ductwork field splice locations.

Figure 10-3 shows some commonly used field splice configurations for rectangular and circular ducts. A wide-flange profile at the splice (detail A) provides good continuity in the duct plate and facilitates the panel-to-panel fit-up.

An E bolt (erection bolt) is shown in the details in Figure 10-3. These bolts allow the erector to pull, fit, and hold adjacent panels together without temporary welds and with minimal temporary attachments. The bolts can be spaced as required, and the structural engineer needs only to designate the bolt material, size, and maximum spacing on the design drawings. The fabricator or detailer will locate and dimension on the shop drawings the erection bolt holes required in the panels. Erection bolts are typically 5/8 in. (16 mm) diameter ASTM A307, with a 13/16 in. (20 mm) diameter oversize hole to permit minor adjustment and alignment of the panels. If erection bolts are used, the design drawings should contain a note to the erector clearly stating that the erection bolt is to facilitate fit-up only and is not to be used to set the critical dimensions of the ductwork. Also, the structural engineer should indicate on the design drawings that after installation the erection bolts should be seal welded on the inside of the duct to prevent leakage around the bolt. An alternative is to have the bolts removed and the holes covered or plug welded.

A typical field splice at duct corners is shown in detail E. As with all splices, the structural engineer should consider the fabrication and erection operations when selecting the corner splice configuration. This corner provides adjustability and additional strength and is typically used on larger ductwork cross sections.

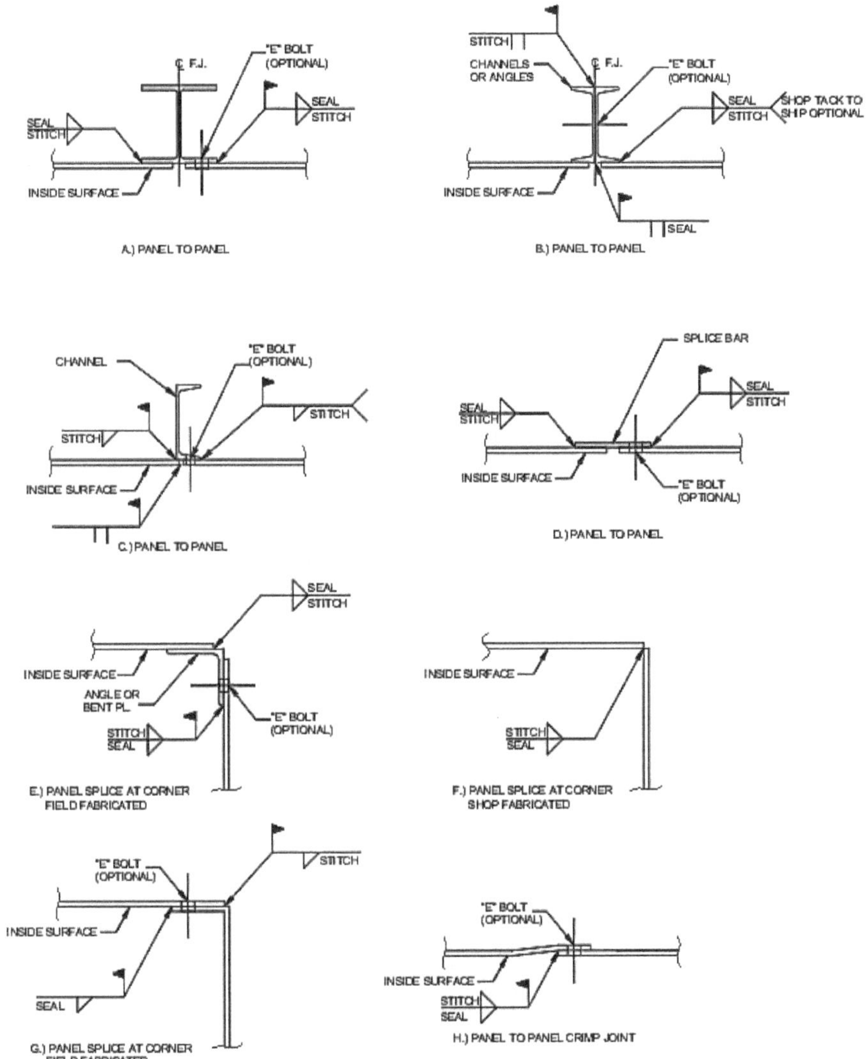

Figure 10-3. Typical field splices for ducts.

Corner splices also play an important role in the transfer of forces in the stiffened panels, as discussed in Chapter 8. For smaller ductwork sections that are shop assembled, detail F is used where it is easier to control fit-up in the fabrication shop.

When joining box sections and circular sections in the field, lap-type joint configurations with unstiffened edges, like the configuration shown in detail H, should be avoided. This is a difficult and time-consuming operation for the erector. Box sections should be joined using a butt-type or flange-type field splice, like those shown in details A, B, C, and D of Figure 10-3. Stiffeners that are shipped loose or only tack welded for shipment may allow field adjustments during erection.

10.2 DRAWINGS AND SPECIFICATIONS

The design drawings, shop drawings, and fabrication specifications must convey all necessary information to the fabrication shop. Erection drawings and erection specifications must convey all necessary information to the field personnel.

10.2.1 Drawing Preparation

Three different ductwork drawing processes are used to convey all the necessary information. The process used is a function of the contract responsibility terms and the preference of the firm responsible for the design.

In the first process, the structural engineer prepares general-arrangement-type drawings, from which the fabricator prepares the ductwork design drawings and the shop fabrication drawings. In this case, the fabricator assumes more design responsibility. In the second process, the structural engineer prepares engineering design drawings that show all supports, stiffener sizes and framing, and truss framing and member sizes, with accurate details and welding symbols. The fabricator may choose to prepare shop fabrication drawings, or he may use the engineer's drawings for fabrication by modifying them as necessary. In the third process, the structural engineer prepares detail-type drawings. In this case, all the necessary information for fabrication and erection is conveyed on one set of drawings, eliminating the need for shop fabrication drawings.

The preparation of ductwork erection drawings is also recommended. These are usually prepared by the same organization that prepares the shop fabrication drawings. Besides piece marks, these drawings can show erection sequence, field fabrication details, lifting restraints, weights, and center of gravity. This information is sometimes shown on drawings, as in the third type of process described above. If erection drawings are not prepared, critical erection information and restrictions must be indicated conspicuously on the design drawings.

Ductwork Design Drawings. The purpose of ductwork design drawings, along with the structural engineer's fabrication specification, is to convey all structural design information to the fabricator. These drawings are created from the plant arrangement drawings and are based on structural engineering calculations performed by the responsible structural engineer.

In the preparation of ductwork engineering design drawings, the following may be considered:

- Usually, at least two views of each duct run are provided. A third view sometimes may be necessary to show a complicated duct section.
- The design drawings should be clear as to whether the dimensions are to the exterior or interior of the plate or shapes.

- The work point of all structural members, especially truss diagonals, should be clearly presented.
- All support details should be clearly shown. This includes slide plates, guide bars, and anchor points. Special care should be taken to clearly show their proper orientation so that thermal growth problems will not occur.
- Construction materials, type, size, and thickness are best shown on each duct section on all drawings.
- All member connection detailing and design information should be clearly indicated. This should include all structural elements and flow distribution devices.
- Specific welding requirements should be clearly indicated, using the welding symbols and procedures specified in the American Welding Society (AWS) *Structural Welding Code, Steel* (2015).
- The structural engineer should consider designing loose expansion joint flanges, cut-to-field-fit sections, and other means to allow easier field fit-up of duct sections.

Ductwork Shop Fabrication Drawings. The purpose of ductwork shop fabrication drawings is to convey all structural design information to the fabrication shop. These drawings are either created directly from the owner's specification and plant arrangement drawings based on structural engineering calculations performed by the fabricator's structural engineer, or created from information in the responsible structural engineer's design drawings and specification.

In the preparation of ductwork shop fabrication drawings, the following may be considered:

- Usually, all views of each duct run are provided. They are usually to scale.
- The drawings should be clear as to whether the dimensions are to the exterior or interior of the plate or shapes.
- Details are best placed on the drawings where they are referenced, as opposed to another drawing.
- Surface preparation and coatings requirements should be clearly marked for the interior and the exterior of each duct section and are best shown on each drawing.
- Construction materials, type, size, and thickness are best shown on each duct section on all drawings.
- Required specific welding requirements should be clearly indicated on the drawings using the welding symbols and procedures specified in AWS (2015).
- Nondestructive examination requirements should be shown on each drawing.
- In general, ductwork fabrication drawings should follow not only accepted industry standards but also the engineer's and the customer's standards.

Ductwork Erection Drawings. The purpose of ductwork erection drawings is to convey all pertinent erection details to the erector. These drawings are created from the ductwork design drawings and the shop fabrication drawings.

In the preparation of erection drawings, the following information may be shown:

- Proposed erection sequence of all duct sections;
- Lifting points and lifting cautions;
- Location of field splices and the field splice details, including all field welding or bolting requirements;
- Piece marks and markings, usually on at least two places per piece;
- Any special field erection tolerances; and
- Duct section or piece weights and center of gravity.

10.2.2 Fabrication Specification

Concise fabrication requirements are necessary to ensure high-quality fabrication of ductwork components. It is not always practical to include all such requirements on the design drawings. Therefore, a document commonly referred to as a fabrication specification should supplement the design drawings. The design drawings should always reference the fabrication specification. They both must be included in the fabrication bid documents and in the erection and construction contracts.

The fabrication specification should include all the fabrication, material, and shipping requirements. It should also reference all the codes and standards applicable to the fabrication. Special nondestructive testing and examination requirements should also be included.

The following is a list of typical information and requirements that should be included in a ductwork fabrication specification.

- Definitions and abbreviations
- Scope of work by the vendor/fabricator
- Required documentation supplied by the fabricator
- Work by the purchaser
- Documentation supplied by the purchaser
- Specific customer/owner requirements
- Applicable codes and standards
- Materials specifications
- Fabrication requirements
- Permissible shop and erection tolerances and shop errors
- Shop connection requirements
- Field connection requirements

- Welding requirements
- Nondestructive testing and examination requirements
- Cleaning and painting
- Quality assurance and inspection
- Weight measurement procedure
- Piece marking and identification
- Shipping and handling requirements
- Shipping information
- Shop and erection drawing requirements
- Document submittal procedures.

10.2.3 Erection Specification

Concise field fabrication and erection requirements are necessary to ensure high-quality field assembly and erection of ductwork components. As with fabrication requirements, it is not always practical to include all such requirements on the design drawings. Therefore, an erection specification may supplement the fabrication specification and the design drawings. The design drawings should reference the erection specification.

The erection specification should include all the field fabrication, erection, unloading, tolerancing, and lay-down requirements. It should also summarize all the codes and standards applicable to the field fabrication and the erection, special field nondestructive testing requirements, and examination requirements.

10.3 FABRICATION

To a large degree, the structural engineer is responsible for the erected product. Specifying fabrication requirements and acceptable fabricated and erected tolerances are ways to help ensure high-quality ductwork installation. Presented here are ways the structural engineer can help accomplish this task.

10.3.1 Plate Fit-Up

The structural engineer can help ensure the squareness of all duct plate by considering plate fit-up when working on the duct arrangement. This is the first step in ensuring that the duct fabrication will conform to the required tolerances.

Plate fit-up should recognize the final required overall dimensions, required gap spacing for welding, and acceptable tolerances. Sufficient temporary internal and external braces should be used by the engineer, the fabricator, or the erector to ensure that fabrication and erection are within the acceptable tolerances. Depending on the size of the ductwork, the structural engineer should assist in the design of any temporary bracing. Shop and field splicing of plates may be performed by many methods, as discussed in Section 10.1.3. Care must be taken in the fabrication process to ensure proper splice preparation for welding and bolting. This includes edge beveling, backing bar attachment, and match drilling of elements. Tolerances should not be accumulated unless special consideration has been made in the design.

Typical Tolerances

- Misalignment between plates at any butt joint should not exceed 33% of the thickness of the thinner plate. When clad plate is used, the maximum misalignment cannot exceed 1/32 in. (1 mm).
- Out-of-square of the rectangular duct cross section should not exceed 1% of the diagonal measurement or 1/2 in. (13 mm), whichever is less.
- Out-of-round of the circular duct cross section should not exceed 1% of duct internal diameter. However, plate mismatch of adjoining pieces should not exceed 33% of the thickness of the thinner plate.
- Peaking of circular duct seams (deviation from a true circle at the seams) shall not exceed 1/4 in. (6 mm) as measured by an 18 in. (0.5 m) long template centered at the weld seam and cut to the prescribed radius.

10.3.2 Fit-Up of Stiffeners, Flanges, and Attachments

Close adherence to design drawing requirements for the attachment of stiffeners, equipment flange connections, and supports is mandatory, because the structural engineer has taken into account many considerations when locating and designing the stiffeners and the duct supports.

Typical Tolerances

- Stiffeners should be within 1/4 in. (6 mm) of the location indicated on the design drawing.
- Equipment connection flanges should be perpendicular to the duct plate within 1/2 degree.
- Support brackets, other ductwork attachments, and openings should be within 1/4 in. (6 mm) of the location indicated on the design drawing.

10.3.3 Assembly of Duct Sections

Shop fit-up of adjacent sections must be carefully monitored. Erection using clear fit-up marks is preferred when possible. When this is not done, an extra total dimension check of all dimensions, considering tolerances, should be performed. The structural engineer should consider designing and showing on the drawings loose expansion joint flanges, cut-to-field-fit sections, and other means to allow easier field fit-up of duct sections.

Typical Tolerances

- For a finished duct section between expansion joints, the total length, width, or height should not deviate by more than 1/4 in. (6 mm) from the dimensions indicated on the design drawings.

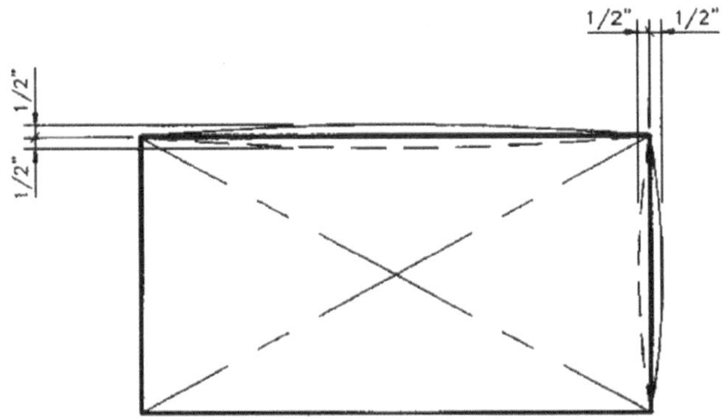

Figure 10-4. Duct cross section bowing, typical tolerance.

- Bowing of completed box duct sections should be limited to 1/2 in. (13 mm), as shown in Figure 10-4.
- For bottom-supported ductwork the difference in elevation between adjacent support foot base plates should be limited to 1/16 in. The difference between two nonadjacent support foot base plates should be limited to 1/8 in.

10.4 WELDING

Of all the operations involved in ductwork construction, perhaps the most critical is the welding. The quality of a completed piece of ductwork can often be judged by the quality of the welding used in constructing the duct section.

Welding processes and methods have improved drastically since the early 1970s. Automated processes are now widely used in fabrication. They provide outstanding quality as well as enhancing productivity. As a rule, ductwork problems associated with welding are rare, thanks to today's advanced welding procedures. However, they do still occur, and unfortunately, they are often not detected in a timely fashion but usually show up after the unit goes operational. This can cause unit operation problems. When ductwork welding problems occur, they are usually widespread and very costly to correct.

Welding is a highly specialized science. Today's welding processes and procedures can be complex. A ductwork structural engineer should not assume the role of a welding engineer. When dealing with unusual materials, extreme service temperatures, and other unusual design parameters that affect welding, experts should be consulted. Some firms have welding or metallurgy experts on staff to assist the duct design engineer in issues pertaining to welding. If welding consultants are available and used, they can prevent expensive welding repairs and the delivering of substandard ductwork to the customer.

However, the ductwork structural engineer needs to have a general knowledge of welding. Some familiarity with welding processes and electrode materials is essential, along with a sound understanding of AWS welding symbols. One of the foremost challenges for the structural engineer is to design accessible welded joints that are structurally adequate and economical to construct. Without knowledge of the welding process, the structural engineer may locate weld joints and configure them in a way that limits access. The resulting welds are usually of poor quality and inefficient.

10.4.1 Welding Codes and Standards

AWS and ASME provide welding requirements that cover the welding of most ductwork. Individual company welding standards are sometimes used to supplement these requirements.

American Welding Society. AWS provides the necessary welding requirements for most ductwork applications. AWS D1.1, *Structural Welding Code, Steel,* is a comprehensive guide for achieving high-quality welding. Ductwork intended to be welded to AWS requirements should specifically reference AWS D1.1 in the ductwork drawing package. AWS D1.3 (2017a), D1.6 (2017b), and A2.4 (2012) may be also applicable to the work.

American Society of Mechanical Engineers. The ASME Boiler and Pressure Vessel Code, Section IX: Welding, Brazing, and Fusing Qualifications, covers the welding of metal products such as ductwork. The welding requirements of Section IX are considered generally more stringent than those of AWS. Many fabrication shops are not certified to do ASME welding. The same is true for many erectors. Thus, the structural engineer choosing ASME Section IX welding should be cognizant of the potential of restricting the choice of fabricators and erectors for the project. This can raise the project cost.

Manufacturer or User Company Welding Standards. Often, specific welding requirements are delineated in the owner's or the ductwork structural engineer's welding standard. Many companies that work extensively in the ductwork field have such standards. These standards supplement the primary welding document, which may be the AWS or ASME code as mentioned above. Where company standards are used, they must be included as a part of the project contract documents issued to the fabricator and erector. A company's welding standard may stipulate any special documentation requirements on a project. For example, welder qualification, weld procedure qualification, and additional weld inspection and testing may be required of the fabricator and erector by the owner. Special nondestructive examination requirements not usually specified may also be in the company's standards. The use of these standards can impact the project cost and schedule, so they should only be used when required and with the permission of the owner.

10.4.2 Types of Welding Processes

A commonly used method of field welding carbon steel ductwork is shielded metal-arc welding, also known as stick-rod welding. To enhance productivity and weld quality, fabrication shops commonly use automated processes such as submerged arc welding, flux cored arc welding, and gas metal arc welding. These processes have specific applications too numerous to mention here. The *AWS*

Welding Handbook, Vol. 2: Welding Processes (2004) introduces and details these various welding processes.

Most of the common welded joints applicable to carbon steel ductwork design are prequalified. By definition they are exempt from testing and qualification per AWS. AWS prequalifies weld joints based on experience that sound weld metal with appropriate mechanical properties can be deposited. Prequalified welded joints should always be used, except where unique conditions prevail. In general, fillet welds are deemed prequalified if they conform to AWS D1.1. Prequalified joints are limited to those made by the manual shielded metal arc, submerged arc, gas metal arc, and flux cored arc welding processes. AWS D1.1 should be referenced for specific welded joint and weld process information.

10.4.3 Welding Electrodes

Many electrode types and sizes can be used for welding ductwork in the various welding processes available. Section 2 of the AISC *Design Guide 21* (2006), illustrates the classification system for electrodes.

The structural engineer need not specify the electrode type to be used for most welding of mild carbon steel. The fabricator and the field erector will select the electrode suitable for the application. But the structural engineer should specify the electrode to be used when welding ductwork joints of materials other than mild carbon steel or when strength is an issue. Often the structural engineer will base the design on a 70 ksi electrode. If this is the case, 70 ksi must be specified so that a 60 ksi electrode is not used.

In addition to selecting the correct electrode for the application, electrode storage and handling are critical to weld joint quality. The electrode material must remain dry and clean before use. Fabricators and erectors should have a welding-consumable procedure as part of their quality program. This procedure should include descriptions of shipping containers, receiving inspection, storage of electrodes and storage containers, baking and redrying, and the disposal of contaminated electrodes.

10.5 SHOP INSPECTION

Periodic shop inspections should be performed by the customer and the structural engineer during the fabrication process. The inspection schedule should be agreed to at the beginning of the work, and it should be maintained. The fabricator's own shop inspection should be an ongoing, daily effort. Some fabricators have implemented total quality management. ISO 9000 quality certification is also fast becoming a method of ensuring high quality in manufacturing processes in the United States. The owner may require the fabricator to be ISO 9000 certified.

At a minimum, the following items should be included in the inspection.

- *Weld quality*. Specified welding requirements (AWS, ASME, or others) must be followed. Industry-accepted standard weld quality inspection procedures for the specified welding methods and procedures should be used. The fabrication specification or ductwork drawings should state any weld inspection criteria required above the minimum criteria given in the applicable welding specification.

- *Dimensions.* Dimensions should be verified. Specified tolerances as shown on the design drawings, in the fabrication contract, and as indicated in the AWS or ASME codes, as applicable, should be used.
- *Material.* Material purchase orders and mill material test reports should be reviewed for compliance with the project specifications.
- *Fabricator's records.* Fabricator's inspection records should be available for review.

10.6 SURFACE PREPARATION

Because most ducts are fabricated from some grade of carbon steel, a surface protection following fabrication is sometimes recommended, regardless of the final specified surface treatment. The surface protection depends on the surface to be protected and the field-applied final coating. This is discussed further in Section 3.7.2.

- *External surfaces.* Degreasing, weld splatter removal, and the general removal of fabrication debris is recommended as the minimum surface preparation. A subsequent coat of primer paint can be applied to deter surface corrosion during shipping and storage. Sometimes it is desirable to shop apply a coating in accordance with Society for Protective Coatings SP 6, *Commercial Blast Cleaning* (1991). Special duct attachments such as instrument port flanges should be greased, coated, covered, or otherwise protected before the duct section leaves the fabricator's shop.
- *Internal surfaces.* In addition to the treatment discussed above for external surfaces, special treatment for internal duct surfaces may be specified for heat or corrosion protection. Supports for linings are often shop applied. Before the duct leaves the shop, the fabricator must take care to protect the internal surfaces and shop surface preparation against possible damage or contamination in transit.
- *Stainless steel.* External and internal surfaces of stainless steel and high-nickel-alloy ducts should be passivated (acid wash followed by neutralizing) in the shop before shipment to clean the surfaces of iron contamination. It is recommended that procedures be implemented in the fabrication shop to eliminate cross contamination of dissimilar metals.

10.7 HANDLING AND SHIPPING

The proper handling and shipping of ductwork is the responsibility of either the fabricator or the erector, depending on the project-specific agreed scope of services. However, the ductwork structural engineer should be aware of proper handling and shipping methods and requirements and traditionally successful methods and procedures.

10.7.1 Piece Marking

The fabricator must clearly mark each piece, assembly, and subassembly that is shipped independently with an erection and shipping mark number. The same mark numbers should be clearly shown on the erection drawings.

Each fabricator should have a written procedure for marking. This procedure may be requested from the fabricator and reviewed by the customer and the structural engineer. The procedure for marking should stipulate that the mark number is tied to the detail drawing number on which the piece is detailed and the section of ductwork that the piece will become a part of. This enables field personnel to quickly tie the piece to its detail drawing and to its duct section for storage and erection sequencing.

Mark numbers should be applied to the components with durable, high-visibility paint in the most conspicuous location. The size of mark should be such that it can be read easily. Preferably, each individual piece will have at least two identical marks on opposite surfaces.

10.7.2 Lifting Lugs

The handling of large ductwork components or completed assemblies usually requires lifting devices. When ductwork is shipped preassembled in relatively large sections, handling can be difficult, and the inclusion of lifting lugs on the duct sections is a necessity.

The ductwork structural engineer may work with the fabricator and erector to determine the requirements for lifting lugs. The equipment and procedures used in handling the large duct pieces are important variables in determining the need for lifting lugs and in sizing the lugs. Without input from the parties doing the actual lifting, the structural engineer risks placing the lugs in the wrong location. However, the design drawings can indicate suggested locations for field-applied lifting lugs or points of lift. Precautions concerning specific inappropriate handling should be clearly noted on the design drawings and in the fabrication specification. A construction impact factor of 1.5 is considered appropriate for design. Alternatively, the contractor responsible for lifting may retain a qualified rigging engineer to design lifting lugs and attachments. The rigging engineer should prepare design calculations for the lifting lugs and for appropriate checks on ductwork strength to support the lifting loads. The rigging engineer should prepare drawings showing the lifting lug design and attachment details. The structural engineer may review the rigging design.

Sizing of lifting lugs is a fairly simple procedure. Chapter D5 of the AISC *Specification for Structural Steel Buildings* (2017) addresses the design of pin connections. It provides a conservative method of determining the size of lugs. The most important steps in the process are to determine the lug material and the design load for the lug. To account for lifting impact effects, the design load should be increased by 50%. Tolbert and Hackett (1974) provides a good explanation of lug behavior and the various stresses associated with lugs.

ASME BTH-1, *Design of Below-the-Hook Lifting Devices* (2017) provides the minimum structural and mechanical design requirements for below-the-hook lifting devices. This standard recommends a nominal design factor for a specific design category based on the predictability of the variation of the loads applied, number of load cycles, and the severity of the environment. This standard does not normally use impact factors. The design factors in ASME BTH-1 are based on load spectra in which peak impact loads are equal to 50% of the maximum lifted load for Design Category B lifters.

10.7.3 Temporary Bracing

The fabricator who ships the ductwork components is also responsible for any temporary bracing required for shipping. Components must be adequately braced during shipment such that they arrive at the project site undamaged and still within tolerance. Site handling also often requires temporary bracing to stabilize the duct section and preserve its shape during erection. The obligation of the fabricator to provide this bracing should be clearly stated in the contract documents in the form of a drawing note or in the fabrication specification. The structural engineer may be called on to assist the erector in determining the bracing needs. ASCE 37, *Design Loads on Structures during Construction* (2015) gives guidance on the types, magnitude, and duration of loads to be used in the evaluation of structures during erection and can be used in the design of temporary bracing.

Circular ducts have a special sensitivity to transportation and handling because of the large potential for plate buckling. When they are to be shipped in one piece as a cylinder, provisions for analyzing and designing them for impact during shipping and handling should be considered. These provisions would apply to the duct supports, bracing elements, and the plate.

Temporary bracing required for shipping should be marked by the fabricator in such a way that the receiving party will clearly know that the bracing is not permanent but is to be removed after erection is completed. One way to accomplish this is by coding temporary pieces with paint marking of a special color. Occasionally, temporary bracing requirements may be shown by the structural engineer on the design drawing for very large pieces or pieces with critical dimensional tolerances. Figure 10-5 shows suggested temporary bracing schemes.

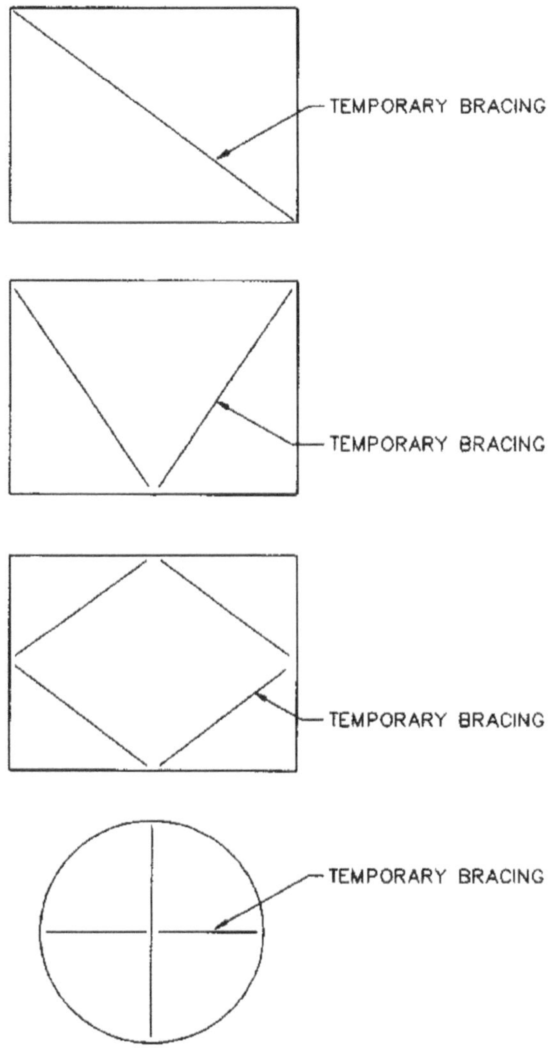

Figure 10-5. Typical temporary bracing schemes.

10.7.4 Loading, Receiving, and Unloading

The shipping party, usually the fabricator, is responsible for loading the ductwork components onto the carrier's transport media. He must supply all cribbing and dunnage necessary to position the items securely on the media. Tie-downs, turnbuckles, and come-alongs are usually supplied by the carrier. If the load is specified to be covered, the carrier must supply the covering.

The carrier's representative, usually the truck driver for truck shipment, must inspect the loaded vehicle and must be satisfied that the load is properly secured. A shipping bill listing all the loaded components should be prepared by the shipper. This shipping bill should always accompany the shipment. When satisfied with the load, the carrier's representative signs the shipping document, and the carrier then assumes responsibility for the load until its arrival at the project site.

Receiving and unloading are the responsibility of the owner, erector, or other authority assigned by the owner to receive the material. On receipt, the receiver should carefully inspect the material. The inspection should include a check that all quantities of material are correct as shown on the shipping bill. If any shortage is identified, the receiver should immediately notify the carrier and the fabricator so that the claim may be investigated.

If materials arrive at the destination in a damaged condition, it is the responsibility of the receiving party to notify the customer, fabricator, and carrier before unloading the material.

10.8 ERECTION

As with fabrication, handling, and shipping requirements, the ductwork structural engineer should be aware of proper erection methods and requirements and traditionally successful methods and procedures. Specifying erection requirements and acceptable erected tolerances are ways to help ensure high-quality ductwork design. Here are some erection considerations for the structural engineer.

10.8.1 Erection Sequencing

Because ducts are designed as structural members spanning support points, proper sequencing for safety should be followed. The structural engineer should note any special erection sequencing requirements and any necessary precautions on the design drawings.

10.8.2 Special Erection Considerations

Temporary Bracing and Supports. The erector should design and furnish any temporary braces or supports required for the erection. This position is taken from Section 7.10 of the AISC *Code of Standard Practice for Steel Buildings and Bridges* (2016). The obligation of the erector to provide this bracing should be clearly stated in the contract documents in the form of a drawing note or in the erection specification. Any temporary erection braces or supports should be removed as soon as practical, and definitely before the duct is insulated and the system goes into service.

The structural engineer may be called on to assist the erector in determining the bracing needs. Occasionally, temporary erection bracing requirements may be shown by the structural engineer on the design drawing for very large pieces or pieces with critical dimensional tolerances. Figure 10-5 shows some suggested temporary bracing schemes.

Field Welding. The specified field weld designation, procedures, and welding material should be followed explicitly by prequalified welders provided by the erector.

Field Modifications. During erection of duct systems, some field modifications sometimes must be made to shop-fabricated components or assemblies because of errors in fabrication or actual field dimensions that differ from the design information. Any field modifications should be approved by the structural engineer responsible for the duct design or the engineer's field representative before the modification is made. The structural engineer may have a good reason to choose an alternate modification in lieu of that chosen by the erector. Any field modifications should be properly documented by the erector in the final job records and ultimately clearly indicated on the engineer's design drawings.

Coating and Lining Repairs. Shop-applied coating and lining should be repaired at field joints and areas of erection damage. Affected areas should be restored by the erector to the originally specified quality.

Inspection. Erected duct systems should be inspected for compliance with the design and fabrication drawings. Verification should include location, dimensions, support clearances, proper bolting and welding, proper attachment of accessories, proper setting of slide bearing plates and guides, and cold-position expansion joint setting. Pressurized duct systems should be inspected for gas tightness by a method such as sonic testing or vacuum box pressure testing using soap or smoke. These testing requirements should be in the erection specification.

Erection Tolerances. The tolerances presented in Section 10.2 should also apply to erection. Tolerances should be nonaccumulated unless special consideration has been given in the design. In addition, the length dimension of an erected duct section should not deviate more than the tolerances specified on the design drawings or identified in the erection specification. In specifying tolerances, the structural engineer should consider that the Fluid Sealing Association's *Technical Handbook for Ducting Systems for Non-Metallic Expansion Joints* (2010) recommends maintaining a 1/4-in. tolerance in the axial dimension and 1/2 in. in the lateral direction.

References

AISC. 2006. *Welded connections– A primer for engineers: AISC design guide 21*. Chicago: AISC.

AISC. 2016. *Code of standard practice for steel buildings and bridges*. AISC 303. Chicago: AISC.

AISC. 2017. *Specification for structural steel buildings*. AISC 360-16. Chicago: AISC.

ASCE. 2015. *Design loads on structures during construction*. ASCE 37-14. New York: ASCE.

ASME. 2015. *ASME boiler and pressure vessel code, Section IX: Welding, brazing, and fusing qualifications*. New York: ASME.

ASME. 2017. *Design of below-the-hook lifting devices: BTH-1-2017*. Washington, DC: ASME.

AWS (American Welding Society). 2004. *AWS welding handbook, Vol. 2: Welding processes*, 9th ed. Miami, FL: AWS.

AWS. 2012. *Standard symbols for welding, brazing, and nondestructive examination*. AWS A2.4. Miami, FL: AWS.

AWS. 2015. *Structural welding code, steel*. AWS D1.1. Miami, FL: AWS.

AWS. 2017a. *Structural welding code, sheet metal*. AWS D1.3. Miami, FL: AWS.

AWS. 2017b. *Structural welding code, stainless steel*. AWS D1.6. Miami, FL: AWS.

Fluid Sealing Association. 2010. *Ducting systems non-metallic expansion joint technical handbook*, 4th ed. Philadelphia: Fluid Sealing Association.

Specialized Carriers and Rigging Association. 2015. *Permit manual*. Fairfax, VA: Specialized Carriers and Rigging Association.

Steel Structures Painting Council. 1991. *Steel structures painting manual, Vol. 2: Systems and specifications. Surface preparation specification no. 6 (SSPC-SP 6), Commercial blast cleaning*. Pittsburgh: Steel Structures Painting Council.

Tolbert, R. N., and R. M. Hackett. 1974. "Experimental investigation of lug stresses and failures." *AISC Eng. J.* **11** (2): 34–37.

CHAPTER 11
Insulation and Lagging

11.1 INTRODUCTION

The design and specification of insulation and lagging is usually the responsibility of the project mechanical engineer. However, there are several aspects of insulation and lagging that have a bearing on the ductwork structural design. It is beneficial for the structural engineer to be familiar with the materials and methods of installation of insulation and lagging.

11.2 PURPOSE OF INSULATION AND LAGGING

Ducts in power plants and similar installations can carry very hot air and flue-gas, with temperatures up to 1,400 °F (760 °C). The purposes of insulation and lagging include thermal efficiency, corrosion prevention, sound attenuation at fan outlets, personnel protection, and keeping water off the hot duct steel. Internal insulation is also used in high-temperature gas turbine exhaust ducts to protect the steel from the metallurgical effects of exposure to the hot gas.

In certain situations, the engineer may decide not to provide insulation and lagging for air ducts. Insulation is often not required for ambient-air ducts because the system temperature is not a design issue for the duct, the flow, or personnel. Also, scrubber outlet ducts are often not insulated and lagged because the flue-gas may be cooler than its dew point, there is typically no need to prevent heat loss, and it allows visual observation of the onset of corrosion. In either case, the engineer should consider protecting the duct steel against the environment with some sort of coating. This is discussed in more detail in Sections 3.7 and 10.6.

11.2.1 Thermal Efficiency

Duct insulation helps maintain the high temperature of air and flue-gas and thus the overall thermal efficiency of the unit. The efficiency of boilers, and thus the unit, partly depends on the temperature of the air and flue-gas in the ductwork system. Therefore, the ductwork should be appropriately insulated so that the minimum amount of heat is lost through the duct system.

11.2.2 Corrosion Prevention

Another purpose of duct insulation is to prevent corrosion of the duct steel. Corrosion can occur when the flue-gas is cooler than its acid dew point, which can be as low as 200 °F (90 °C). Corrosion occurs when water vapor in the flue-gas combines with sulfur from the coal or oil ash to form sulfuric acid. Over time, this acid will corrode the duct casing. The duct steel plate temperature should be kept above the dew point of the flue-gas unless the steel duct plate is properly protected or fabricated from a corrosion-resistant material. Maintaining a temperature above the acid dew point within the duct is an important function of the insulation.

11.2.3 Personnel Protection

Insulation and lagging provide a barrier for personnel protection. The exposed exterior of ducts should be kept at a temperature low enough to be safe for plant workers to touch. This temperature is specified as 140 °F (60 °C) in ASTM C1055, *Standard Guide for Heating System Surface Conditions That Produce Contact Burn Injuries* (2014). The temperature limit should be discussed with client and plant personnel, particularly at insulation and lagging protrusions such as duct supports and instrument ports where the temperature may exceed 140 °F locally, to determine whether additional safety guards are warranted.

11.2.4 Protection against Water

The greatest external cause of damage to duct steel is prolonged exposure to water. Very hot duct plate can buckle and crack from thermal shock if exposed to cool water on one side and hot flue-gas on the other side. Standing water on the top of a duct can also cause accelerated corrosion by dropping the temperature locally below the flue-gas acid dew point. If the insulation gets wet, it can act like a sponge and retain water for a long time, and wet mineral wool insulation may compress and lose some thermal efficiency. This only further accelerates the corrosion process. Keeping water away from the duct steel and the insulation is the most important function of the lagging.

11.2.5 Reduction of Steel Temperature in Very Hot Ducts

As discussed in Chapter 3, there are significant effects on carbon steel that should be considered when it is to be exposed to high temperatures. These effects can raise the cost of the ductwork considerably. To make carbon steel an economic option (compared to stainless steel or alloys) at very high temperatures, internal insulation can be used to protect the duct steel from the metallurgical effects of exposure to the hot gas. This insulation method lowers the duct steel design temperature to a reasonable level so that the steel's performance will not be degraded. This approach is typically used in high-temperature gas turbine exhaust ducts.

11.2.6 Sound Attenuation

The large fans of power stations and industrial boilers can be very noisy. Silencers may be located on the inlet side of some fans. Less frequently, mufflers are installed in fan outlet ductwork; however, the pressure losses associated with mufflers usually make their use unacceptable or costly. Insulation and lagging will provide some reduction of the sound of fans coming through the ductwork. Project noise studies are sometimes done to assess the type of sound attenuation that may be needed, such as inlet silencers, outlet mufflers, and/or specially designed sound-attenuating insulation and lagging.

11.3 TYPES OF INSULATION AND LAGGING

Insulation may be applied to the inside or outside of ductwork. It must be asbestos free. More recently utilities have also been requiring that insulation be free of ceramic fibers. Today, externally applied insulation covered with lagging is most commonly used.

11.3.1 Insulation

A multitude of materials are used as external duct insulation. The specific manufacturer's recommendations for temperature limits and thicknesses should be followed when specifying an insulation system. The most commonly used insulation materials are as follows.

Inorganic glass fiber is an external insulation normally used for temperatures up to 850 °F (450 °C). Its temperature range depends on its resinous binders. The material is supplied in various sizes in semirigid boards or blankets from 1 in. to 4 in. (25 mm to 100 mm) thick.

Noncombustion wool insulation (mineral wool) is produced from mineral fibers. It is an external insulation that can be used for temperatures up to 1,200 °F (650 °C). It is supplied in various size batts from 1 in. to 4 in. (25 mm to 100 mm) thick.

Insulating mortar. Mortar or sand aggregate concrete insulation can be applied to the internal duct plate by a method commonly known as shotcrete or gunite. The standard practice is to apply a 2 in. to 6 in. (50 mm to 150 mm) thick layer of mortar over wire mesh. This reinforced mortar is anchored to the duct by studs welded to the plate. This insulation method would normally only be used today at the client's request, to match an existing condition, or to protect the duct steel plate against harsh flue-gas. There are several reasons it is no longer common. It is much heavier than fiberglass or mineral wool and is more expensive to install. This adds to the cost of the duct and its support structure. History has shown that internal mortar has a tendency to crack and will not always adequately protect the duct plate, because moisture and acid can be trapped between the mortar and duct plate, contributing to corrosion after only a few years of service. However, manufacturers have recently developed protective membranes that are spray applied to the duct plate and then top coated with mortar by shotcreting or guniting. This method provides steel corrosion protection plus insulating value, and it protects the protective membrane from abrasion. Manufacturers have also developed chemical-resistant insulation systems that are cementitious and gunitable and do not require wire mesh reinforcement. These systems typically use specially designed anchors, such as T anchors, to reinforce and support the gunite. After the anchors are installed, a chemical-resistant membrane is applied to the steel plate and anchors before the cementitious lining is installed.

Ceramic fiber insulation is often applied internally, typically in very high-temperature gas turbine exhaust ducts. It is supplied in various thicknesses up to 6 in. (150 mm). Ceramic fiber is usually covered by thin-gauge stainless steel internal lagging to protect it from the flue-gas. The stainless steel lagging is attached in such a manner that it can expand with increasing temperature. A rigidizer coating is available to resist erosion of the ceramic fiber. The use of ceramic fibers is being phased out by some utilities in view of respiratory concerns related to inhalation. The engineer designing the insulation system should confirm whether the use of ceramic fibers is permitted on the project site before specifying them for use on a project.

Refractory lining is a high-temperature internal lining that is available for use in ducts with operating temperatures above 1,000 °F (540 °C). It may also be used where a gas or oil-fired burner is installed inside a duct or in circulating fluidized-bed gas ducts. It is not commonly used in lower-temperature ductwork.

Borosilicate block internal insulation provides an acid-resistant surface along with insulating value. It is installed similarly to a brick or refractory lining. When using this material for acid protection, the proper acid-resistant adhesive must also be used.

11.3.2 Lagging

Lagging is the covering used to protect insulation. External lagging protects the insulation from weather, especially wind and rain. External lagging must be resistant to weather and is typically constructed with corrugated panels that overlap at least one full rib to provide a weather-tight connection. Internal lagging must protect the insulation from the abrasive effect of the high-velocity flue-gas. Lightweight material is usually used, to reduce the overall duct weight and the cost of the lagging, and for ease of construction. The most common materials are aluminum, fiberglass-reinforced plastic, galvanized or aluminized carbon steel, and stainless steel. Lagging is commonly supplied in corrugated sheets to provide some rigidity in handling and so it can more efficiently carry externally applied loads such as lay-down loads, wind, and foot traffic. For ease of construction, flat sheets of lagging are sometimes used for circular or irregularly shaped ducts and in areas of tight clearance. Roof lagging may be thicker than wall lagging because it should be designed to withstand foot traffic from maintenance personnel without excessive deflection.

11.3.3 Subgirts and Standoffs

Subgirts and standoffs are usually custom-fabricated out of sheet metal as required for the installation of the insulation and lagging system. The material is usually aluminized structural steel or Alclad aluminum. The designer should choose materials that are compatible with the lagging and will not cause galvanic corrosion. If dissimilar metals are in contact, they should be separated by insulation tape or other means to prevent galvanic action. With aluminized steel subgirts, the attachment of the subgirt to the duct stiffeners is not typically a concern for galvanic action, because should the aluminum coating deteriorate, the carbon steel underneath would be a similar metal to that of the duct stiffener.

11.4 EFFECTS OF INSULATION AND LAGGING ON THE STRUCTURAL DESIGN OF DUCTS

The type and thickness of insulation and lagging should be determined early in the ductwork design process. The insulation should be chosen by the mechanical process engineer. Both the mechanical process engineer and the structural engineer should have input on the design of the lagging. Information regarding the insulation, mainly the dead weight, affects certain aspects of the structural design of the ductwork.

11.4.1 Layout Considerations

The sizing, layout, and routing of a duct system must take into consideration the projected thickness of the insulation and lagging all around the duct to allow proper installation of the insulation and lagging without any interference from adjoining equipment or structural steel. The top of horizontal duct lagging surfaces should be pitched to divert rainwater. Gutters and downspouts attached to the duct lagging are also sometimes used to channel water away to proper areas. Special attention should be given to providing proper clearances for the sloped roof insulation and lagging. Thermal movements must also be included in considering clearance from adjacent fixed structures to the insulation and lagging, which move with the duct.

11.4.2 Duct Design Considerations

In the early stages of the duct layout and structural design, the structural engineer should obtain from the mechanical engineer the type and thickness of insulation and lagging being specified for the ductwork. Ideally, if the ducts have external stiffeners, the structural engineer should size the stiffeners on each duct panel to be of a uniform depth, if an economical duct design can be achieved. This will make it easier and less costly to install a uniform thickness of insulation and lagging over the stiffeners.

The project team may request that the structural engineer specify the thickness and profile of the lagging. The structural engineer would then design the lagging for the appropriate spans, the expected live load, and the proper snow and wind loads. More typically, the loading requirements are included in the lagging procurement specification, and the thickness, profile, and support requirements are determined by the insulation contractor.

11.4.3 Dead Load Determination

The structural engineer should include in the dead load calculations the weight of insulation, lagging, and their attaching systems, if any. This attaching system is usually referred to as subgirt framing or stand-offs. The average density of fiberglass insulation varies from 3 pcf to 6 pcf (48 kg/m^3 to 96 kg/m^3), whereas mineral wool may go up to 12 pcf (190 kg/m^3). Lagging weight depends on the thickness of the material, ranging from 0.5 psf to 1.0 psf (24 kg/m^2 to 48 kg/m^2) for aluminum, 0.5 psf to 0.75 psf (24 kg/m^2 to 36 kg/m^2) for fiberglass, and 1.5 psf to 3 psf (72 kg/m^2 to 144 kg/m^2) for steel. (The weight of the attaching system is included in these values.) Should mortar or sand aggregate concrete be used as internal insulation, the approximate weight is 12.5 psf per inch of thickness (235 kg/m^2 per cm). Borosilicate block weighs 1.5 psf per inch of thickness (30 kg/m^2 per cm). The insulation and lagging weight assumed in the initial design should be verified based on the actual procured material.

11.4.4 Wind Load Determination

Calculation of the wind load on the ductwork sections should be based on the overall outside dimensions of the duct, plus external stiffeners, plus the insulation and lagging. This additional width of insulation and lagging outside the stiffeners can vary from 2 in. (50 mm) to as much as 9 in. (230 mm), depending on the stiffener arrangement, the insulation and lagging system, and the slope of the insulation and lagging at the roof. When installing these cantilevered sections of insulation and lagging, the contractor must make sure that the wind conditions are considered.

The local wind effects, as stipulated under "Components & Cladding Wind Load Provisions" in ASCE 7-16 (2017), *Minimum Design Loads and Associated Criteria for Buildings and Other Structures,* should be considered when designing the lagging, subgirts, and attachments. If the lagging is designed by the supplier, the structural engineer should specify this requirement in the procurement.

11.5 METHODS OF INSTALLATION AND QUALITY OF THE WORK

Insulation and lagging installation methods and details are, in general, determined by the insulation and lagging supplier or contractor. The design of the insulation and lagging subgirts, lagging, and so on should conform to the American Iron and Steel Institute's *North American Specification for the Design of Cold-Formed Steel Structural Members*, S100 (2012) or other appropriate industry codes. The structural engineer should review the contractor's installation methods and details, because the insulation and lagging are important in protecting the duct steel. Any failure of the insulation and lagging to protect the duct could reduce the performance of the duct steel and the structural integrity of the ductwork.

11.5.1 Construction Details

When the installation details are prepared by the contractor, there are three important installation requirements that the structural engineer should specify.

- A continuous and uniform thickness of insulation should blanket the entire duct exterior in accordance with the specification and proper installation practice.
- The insulation should cover the entire exterior face of all steel stiffeners so that minimum thermal gradients will be created.
- The insulation around expansion joints and dampers and their associated frames should be installed in accordance with the manufacturer's recommendations or requirements.

11.5.2 Installation Methods

There are many methods for installation of insulation and lagging. Some typical methods are discussed here. Some of these methods may be used for either shop or field installation. These are only illustrative in nature; many contractors may have developed their own acceptable methods.

External insulation applied directly to the duct plate. The installation of external insulation directly on the duct plate is shown in Figure 11-1. This method is ideally provided when internal duct stiffeners are used. Installation is usually performed in the following sequence.

1. Metal pins are welded directly to the duct plate.
2. Insulation boards are impaled over the metal pins and held in place with metal clips, sometimes called speed clips, which fasten onto the pins. If blanket-type insulation is used, it is usually held in place over the pins with a layer of wire mesh.

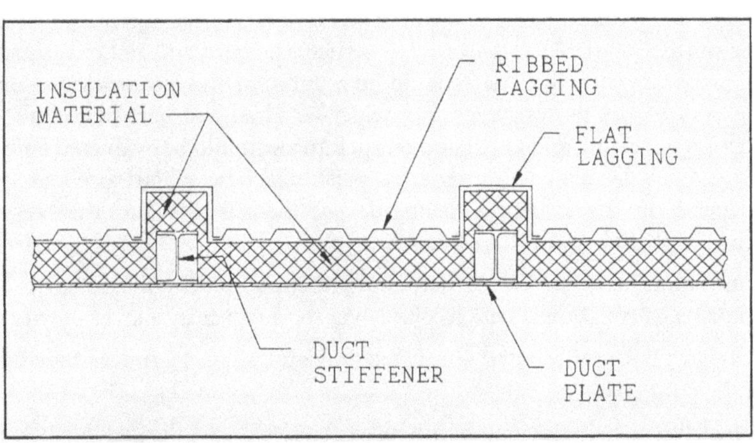

Figure 11-1. Typical external insulation attached to the duct plate.

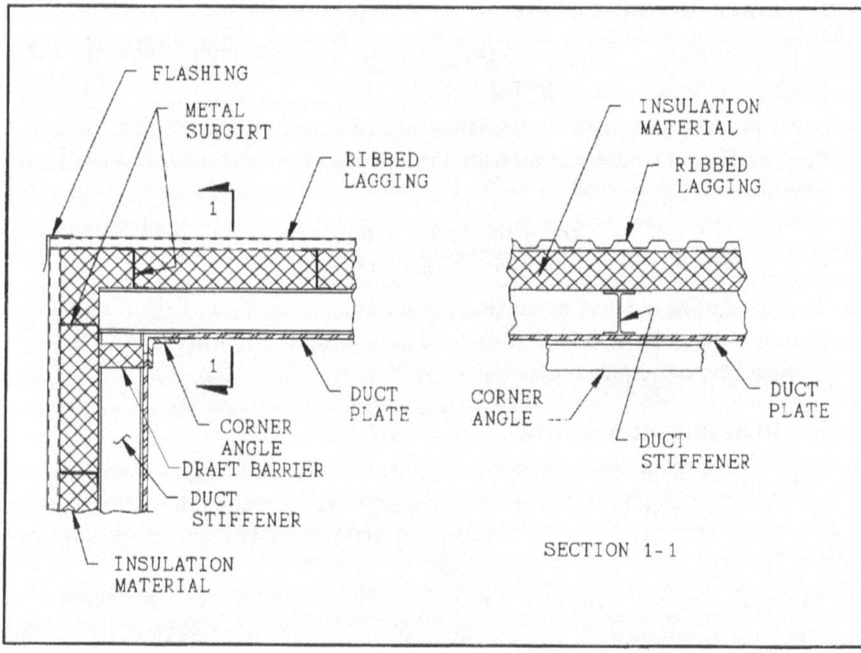

Figure 11-2. Typical external insulation attached to duct stiffeners.

3. Subgirts are attached to the clips.
4. Lagging is applied over the insulation and fastened to the subgirts.

External insulation applied to a subgirt system. This method is shown in Figure 11-2. When the insulation and lagging are installed onto a subgirt system over the external stiffeners, installation is usually performed in the following sequence.

1. Light-gauge metal subgirts are attached to the external stiffeners by welding or powder-actuated fasteners. The structural shear and pullout capacity of powder-actuated fasteners should be reduced as the ductwork temperature increases. The performance of powder-actuated fasteners at the ductwork design operating and excursion temperatures should be evaluated before the use of powder-actuated fasteners is permitted. The subgirt size and spacing depend on the structural properties of the subgirts and the lagging in resisting wind, snow, and live loads.
2. Insulation batts are placed between and supported by the subgirts. Sometimes the insulation is held in place over the subgirts by a layer of wire mesh.
3. Lagging is installed over the insulation and subgirts and fastened to the subgirts, usually by screws.

When this installation method is used for hot ducts, the space between the duct side wall and the insulation allows a "chimney effect" up the sides of the duct that may drive the temperature of the lagging at the top of the duct beyond the specification requirements. This effect could also cause hot air to escape from the lagging at the top edges of the duct, which could be a hazard to personnel. And cold air could be sucked into the dead space between the duct plate and the insulation, causing a cold spot on the duct plate, which would subsequently corrode. To prevent the chimney effect, draft barriers consisting of blocks of insulation are typically installed between the duct plate and the insulation along the duct corners and at an intermittent spacing along the vertical surfaces of ductwork.

Internal mortar or refractory insulation. The use of mortar or refractory insulation is shown in Figure 11-3. When the mortar insulation is installed internally, it is usually installed onto the duct plate in the following sequence.

1. Studs or anchors are welded to the interior duct plate.
2. The inside surface of the duct plate is sandblasted.
3. A protective membrane is applied.

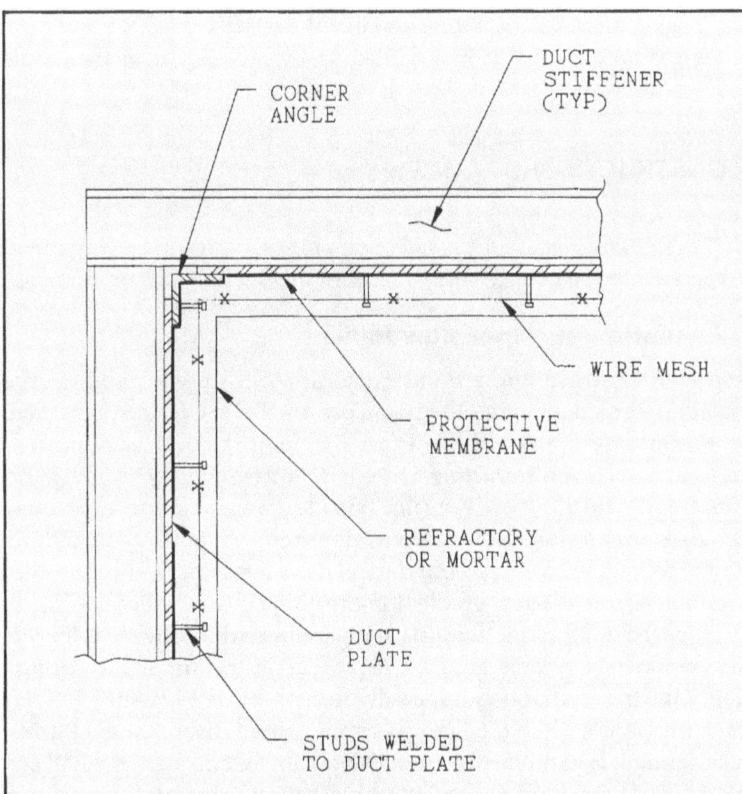

Figure 11-3. Typical internal insulation details.

4. Wire mesh may be installed to reinforce the refractory; however, some manufacturers have developed specialized anchors that are designed to both reinforce and support the mortar and thus eliminate the need for the wire mesh.
5. Cementitious refractory or sand aggregate mortar is applied to the surface to a depth of 2 in. to 6 in. (50 mm to 150 mm).

Internal insulation applied directly to the duct plate. This installation method is like that shown in Figure 11-1 except that the insulation is applied to the inside of the duct and the lagging is usually flat and thicker than external lagging. When internal insulation and lagging are attached directly to the duct plate, the installation is usually performed in the following sequence.

1. Metal pins or studs are welded directly to the duct plate.
2. Insulation is impaled over the pins and held in place with metal clips that fasten onto the pins. The insulation is sometimes held in place over the pins with a layer of wire mesh.
3. Lagging is applied over the insulation and fastened to the clips. This must be done in a manner such that the lagging can expand as the gas temperature rises. The lagging must also be installed so that the seams are lapped in the direction of the gas flow, so that it is not lifted up by the velocity of the gas flow.

11.6 CONSTRUCTION DETAILS

There are several areas where the installation of insulation and lagging can have an adverse effect on the structural components of the ductwork if not done properly.

11.6.1 Insulation at Expansion Joints

Insulation and lagging at and around expansion joints poses a unique problem. The insulation should be continued as close to the expansion joint flanges as possible without interfering with the expansion joint removal, as discussed in the Fluid Sealing Association's *Ducting Systems Non-Metallic Expansion Joint Technical Handbook* (2010). However, the outside flanges of the expansion joint channel frames may be left uninsulated to allow removal of the joint, although the portion of the channel closest to the duct plate is insulated. With hot ducts, this sets up an undesirable thermal gradient in the frame channels because the interior flange is exposed to the hot flue-gas temperature while the exterior flange is exposed to ambient air. This gradient can cause distortion and cracking in the expansion joint frame channels, especially near their connections. This problem predominantly occurs when the flue-gas temperature is over 300 °F (150 °C). One way of alleviating it is to modify the expansion joint and insulation detail as shown in Figure 11-4. The proposed insulation method at the expansion joint frames should be approved by the expansion joint supplier.

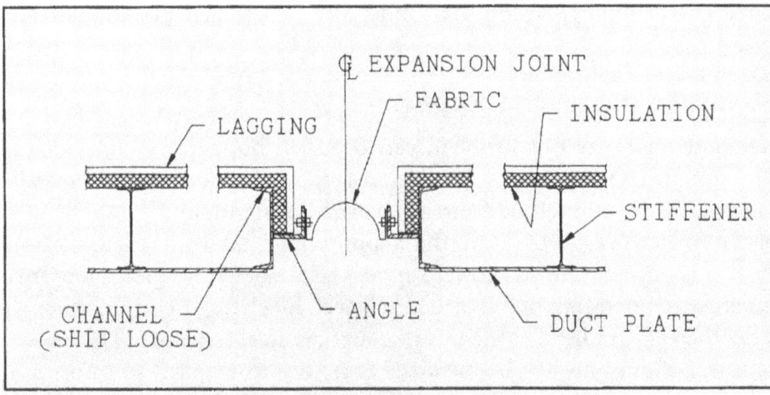

Figure 11-4. Typical insulation detail at expansion joints.

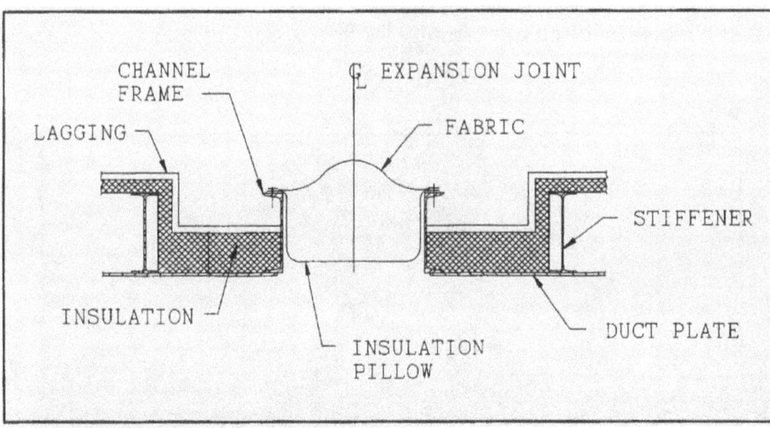

Figure 11-5. Alternate insulation detail at expansion joints.

For air ducts, a detail similar to Figure 11-5 may be used. This detail can also be used for cold-side flue-gas ducts if the duct and the expansion joint frame are sufficiently flexible to accommodate the rotations caused by the thermal gradients that are set up in the expansion joint frame.

The expansion joint vendor should provide recommendations for insulating around and over the expansion joints. Typically, vendors will not want to insulate over expansion joints on hot-side ductwork because the insulation will trap in heat, which can cause the expansion joint materials to deteriorate. On cold-side ductwork, vendors may recommend insulating over the expansion joints to keep the material warm and to prevent cold spots from forming that could lead to the condensation of sulfuric acid. Acid could degrade the expansion joint materials and/or the metal expansion joint frame. When insulating over expansion joints, compressible insulation materials should be used to allow for thermal movement.

The lagging over the expansion joint must also be detailed to allow for thermal movement.

11.6.2 Insulation around Hoppers

Another potentially serious problem can occur in hopper crotches adjacent to exterior support steel if the insulation and lagging are not properly installed in these areas. The easiest method from a construction standpoint would be to install the insulation and lagging below the support steel between adjacent hoppers. However, this encases the structural support steel inside the insulation envelope and raises the temperature of this structural steel. Insulating in this manner will set up a large thermal gradient in the exterior support steel structure. This potentially serious effect should always be avoided. One way to avoid it is to use a detail similar to Figure 11-6.

11.6.3 Insulation around Duct Appurtenances

Duct appurtenances, such as access doors, flow monitoring ports, dampers, and expansion joints provide an opportunity for heat loss, rainwater intrusion, and the

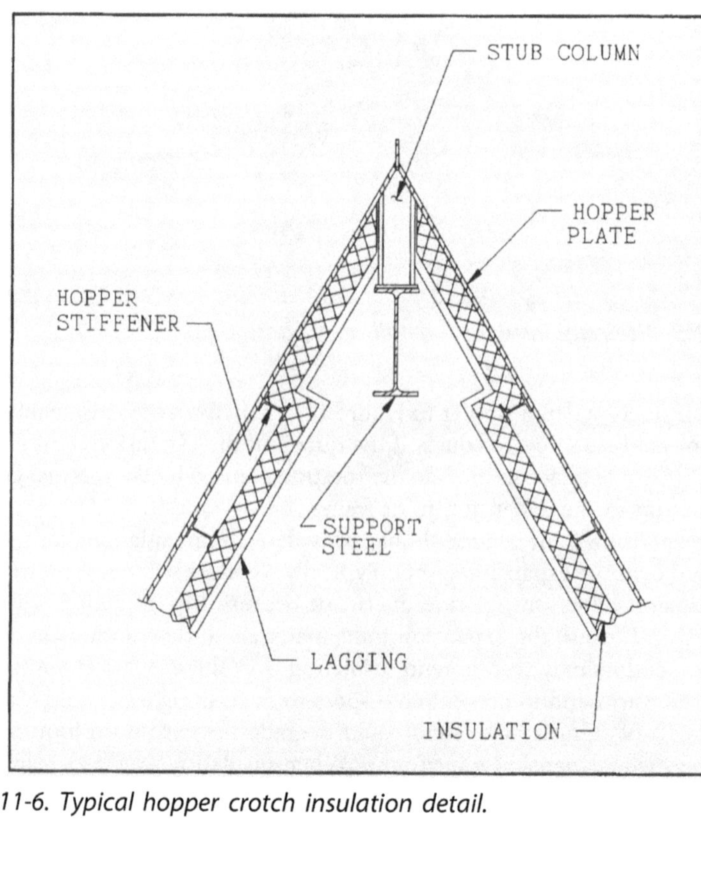

Figure 11-6. Typical hopper crotch insulation detail.

accompanying metal corrosion. Access doors may be preinsulated by the manufacturer before shipping. However, if field-fabricated or flush-mounted doors are being used, a removable piece of insulation and lagging should always completely cover the door. This removable piece should always be securely attached with bolts. The installer should confirm with the door manufacturer that covering the door in this way will not reduce the door's performance.

Insulation and lagging around a louver damper should be installed to allow clear access to the actuating linkage and shaft bearings. This permits maintenance to be performed without removing the insulation and lagging and also avoids possible binding of the insulation in the damper linkages when the damper is opened or closed. The proposed insulation method at the damper should be approved by the damper supplier.

11.6.4 Attachment of Lagging

Lagging should always be firmly attached to its supports. It should be laid out with sufficient overlap to prevent intrusion of water. This will also prevent wind from getting under the sheets and lifting them off. All edges and corners should be covered with flashing, for the same reasons. Lagging on the top surfaces of ducts, including at the expansion joint frames, should always be sloped so that rainwater will drain away. When this is done, the seams must be lapped so that the water will flow over the seam, not under the seam and onto the insulation and duct plate. At the edge of the duct at the bottom of the slope, there must not be any flashing dams that will stop the water from flowing over the edge of the duct. All of these drainage details are important because the top duct plate and external stiffeners should never be exposed to standing water.

In areas of tight clearance, the insulation and lagging system may have to be locally modified. Flat sheets of lagging may be used in these areas instead of ribbed material. Also, if the specified method is external insulation attached to subgirts, in areas of tight clearance the engineer may elect to locally place the insulation directly against the duct plate, as shown in Figure 11-1, to save space or to make the installation easier.

11.6.5 Attachment of Subgirts

When attaching the subgirts, an allowance must be made for the thermal growth of the duct in relationship to the much cooler lagging. This is particularly true when the lagging spans expansion joints. In this case the lagging must be designed and constructed to accommodate all of the expansion joint's design movements.

References

AISI (American Iron and Steel Institute). 2012. *North American specification for the design of cold-formed steel structural members: S100-12*. Washington, DC: AISI.

ASCE. 2017. *Minimum design loads and associated criteria for building and other structures*. ASCE 7-16. Reston, VA: ASCE.

ASTM International. 2014. *Standard guide for heated system surface conditions that produce contact burn injuries.* ASTM C1055. West Conshohocken, PA: ASTM.

Fluid Sealing Association. 2010. *Ducting systems non-metallic expansion joint technical handbook*, 4th ed. Philadelphia: Fluid Sealing Association.

CHAPTER 12

Maintenance Examination of Existing Duct Systems

12.1 FACTORS THAT INFLUENCE THE NEED FOR STRUCTURAL EXAMINATIONS

Ductwork systems are often exposed to adverse chemical conditions and high temperatures. Many ducts are also subject to varying environmental conditions associated with local weather and atmospheric pollution. The purpose of preventive maintenance examinations of duct systems is to ascertain the structural integrity of the ducts to function safely and efficiently as originally designed after repeated exposure to these adverse conditions.

Owners and plant operators may consider ducts permanent structures with virtually unlimited strength and durability. Experience has shown this belief to be generally untrue. It is common for a design to be based on a particular design life, and design decisions may have been made in material selection and/or corrosion and erosion allowances based on the selected design life. The actual operating conditions may or may not match the exposure conditions assumed in the original duct design, and in many situations the ducts have been in operation past the original design life. Regular ductwork structural examinations should be part of a planned preventive maintenance program that would prevent structural deterioration and costly repair or replacement of the ducts. Therefore, such structural examinations are just as important for ducts as normal scheduled examinations and maintenance are for other plant equipment. Unfortunately, given budgetary considerations, it is sometimes difficult for owners to include such regularly scheduled examinations by competent ductwork structural engineers in their maintenance programs. Owner inspections by plant maintenance personnel will occasionally not recognize the structural significance of their observations.

Selection of and consistent use of common terminology at the start of a structural maintenance examination is beneficial for all involved. An understanding of the duct system definitions will contribute to successful communication of examination tasks, procedures, data, and reports.

12.1.1 Original Design Basis versus Operating Basis

Both the original design and the historic operating basis for the duct system should be reviewed before any comprehensive examination. The objective of the review is to fully understand the service conditions assumed by the original design engineer, as well as the conditions under which the duct system has been operating. A comparison of the two bases can be informative and can help identify potential problems or explain observed problems.

Items to be reviewed may include the design input or criteria, design calculations, and design drawings for the original system design. Further attention should be given to available shop fabrication details and erection drawings, previous inspection reports, ongoing maintenance reports, and operating data. Operating data could include normal occurrences such as operating temperatures and pressures, as well as unusual events that have occurred during operation.

The structural engineer should become aware of any modifications to the ductwork system, support structures, or mode of operation in the system's operating history. Equipment additions or changes may cause the redistribution of loads to a structure originally intended to support only the duct. Similarly, modification or capping of ash collection hoppers may change the fly ash fallout, resulting in more or denser ash in the downstream ducts. Revised plant or boiler operating conditions may result in the ducts being exposed to higher pressures, higher temperatures, thermal gradients, or fly ash fallout. Changing to a different fuel source may produce higher loads from the accumulation of ash or sludge, or accelerate the rates of erosion or corrosion.

Though most duct system designs incorporate a certain degree of conservatism, the accumulated effect of these modifications could present a safety concern. Areas that exhibit differences between the original design basis and the current operating basis warrant a more detailed structural examination.

12.1.2 Exposure to Weather, Chemical, and Thermal Conditions

Outdoor ducts may incur damage from the accumulation of snow and ice or the variable loads associated with gusting wind and driving rain. Snow drifts may melt and subsequently refreeze, expand within the insulation, and damage the insulation and lagging. This damage could provide paths for further water intrusion. Water from frequent dampness, rain, or melting snow also may find a path through seams in the lagging to the underlying insulation, duct plate, stiffeners, and expansion joints, creating cold spots where the localized temperature may dip below the acid dew point, and accelerating duct corrosion. Plant wash-downs can transport plant by-products and water onto the duct. The process of deterioration will begin and steadily progress once water or plant by-products penetrate the duct lagging into the undrained areas between stiffeners or at expansion joints. The chemistry of the accumulated water may also be altered by plant by-products, creating acidic conditions that may accelerate the damage to the steel duct or insulation.

Insulated ducts may experience thermal variations, commonly known as cold spots, at areas of damaged insulation, areas of external water or air contact, expansion joints, test ports, supports, and platform penetrations. The cooler duct surface may cause internal condensation and eventual deterioration.

The impact of FGD processes on the integrity of a duct system is of particular importance because of the highly corrosive nature of the by-products created and collected by such processes. FGD uses a variety of additives to condition the flue-gas and to enhance the removal of pollution particles. These additives may also contribute to the internal degradation of the duct. Progressive interior deterioration can result in holes in the duct plate, which will cause inward leakage, expose plant personnel to hazards, and threaten both internal and external structural elements and the overall structural integrity of the duct and nearby structural components.

Thermal growth presents numerous opportunities for the development of structural damage. Such growth, whether uniform or distorted, produces friction forces in bottom-supported ducts that must be resisted by the supporting structure. Original designs typically anchor a duct at one location and guide the system's thermal growth to other locations, where slide bearing plates are usually provided. Progressive deterioration or misalignment can increase these friction forces. Thus, any examination should include an assessment of the capability of the duct support legs, the support steel, and the slide plates to transmit thermally induced friction forces and resist associated thermal deformations.

12.1.3 Duct Erosion

Upgrades or changes to internal flow distribution devices such as splitter plates and turning vanes may increase the dead load, alter the duct system's vibrational characteristics, or alter localized flow velocities. Locations of high flow velocities may be prone to erosion, and any erosion of the internal bracing and the respective connections is of critical importance. The rate of erosion depends on several factors, including the flow velocity, profile, and duration, the chemical composition of the fly ash, and the steel hardness. Many of these factors may be difficult to quantify to the accuracy necessary to determine the actual rate of erosion. High-velocity-flow regions should be carefully examined to verify that there is sufficient material for structural integrity and to evaluate whether the rate of erosion is increasing or decreasing within the duct segment to estimate the life remaining before repairs or replacement will be needed.

Before a ductwork examination, it is best to know what changes may have been made to the system or operation that could impact the flow characteristics. For example, if a plant has switched to a different type of coal that results in a higher percentage of silica in the fly ash or has increased the fan operation, more erosion might be observed in the internal duct elements. Or the plant may have removed perforated plates that were originally installed inside the ductwork because they were eroding away, but this may result in less flow stratification and a higher concentration of localized erosion on internal elements.

12.1.4 Operations Records

Many plants maintain extensive records of ductwork structural problems, behavior problems, damage, and associated structural repairs. Repairs of duct accessories, such as expansion joints, dampers, and control linkages, may indicate a problem in the ductwork structure or its arrangement. Also, inefficiency of the fans may be attributed to extensive inward leakage of air caused by plate deterioration. When records indicate that frequent outages of significant duration have been necessary for the repair of ductwork, a comprehensive maintenance examination should be scheduled.

Plant records may also identify unusual events of major or minor significance to the structural engineer. From a structural standpoint, information pertaining to the magnitude, duration, and frequency of transient pressure and excursion temperatures may provide a basis for review or resolution of ductwork structural concerns. Ideally, an examination is performed immediately following a transient pressure or temperature event to ensure that significant damage has not occurred. Any information regarding past fires in the ducts may also be significant. Knowledge of chronic large ash or sludge deposits in sections of the ductwork is also helpful when planning the examination and when reviewing and evaluating the collected examination data.

The lead examining structural engineer should also preferably interview the plant staff before starting the examination to identify potential problem areas. Historical concerns or ongoing chronic problems are most often clearly presented by plant operators, site-based results engineers, and long-term maintenance staff.

12.1.5 Cumulative Effect of Exposure and Operating Conditions

As mentioned in Section 12.1.2, rain, snow, ice, wash-down water, and plant by-products can provide the catalyst for accelerated duct deterioration. Frequent exposure to water may deposit acidic concentrations that can accelerate the corrosion of the duct plate and external stiffeners. In hot ducts, cold spots may induce thermal shock, metallurgical changes in the steel (quenching), corrosion, cracking, and subsequent deterioration. Such cumulative effects often develop within ductwork systems that are externally lagged and insulated. Yet they can go unnoticed, because leakage at lagging seams may not be obvious from outside.

The cumulative effects of plant operations also may influence ongoing structural deterioration. Repetitive exposure to such unusual conditions as transient pressures and excursion temperatures may induce mechanical damage or accelerate material deterioration. Unexpected thermal gradients can also cause damage. These conditions, when combined with cyclic plant operations, can result in the binding of sliding surfaces and damage to the critical interfaces with the support structure. Thus, forces and events not necessarily considered in the original design may induce buckling and tearing of the duct plates and structural elements of the ducts.

12.1.6 Ensuring Structural Integrity

A structural preventive maintenance examination should assess the cumulative effect of plant operations on the structural integrity of a duct. Plant downtime and unit outages for structural repairs are obviously not desirable. Simply stated, the function of ducts is the movement of air and flue-gas. Many potential problems that can jeopardize this function, as well as the safety of plant personnel and the duct's structural integrity, can be identified and corrected with scheduled preventive maintenance examinations. The scheduling may be constrained by the plant operating cycle, so plant outages may not coincide with desired monitoring frequencies. Still, structural preventive maintenance examinations of ductwork should be part of regular plant maintenance.

These routine condition assessments should also be considered an integral part of all safety-related and plant life extension programs.

12.2 FIELD EXAMINATION TECHNIQUES

To facilitate both exterior and interior ductwork examination, sufficient time should be invested in adequate preparation and pre-outage planning. Based on the importance of the factors presented in Section 12.1, the following types and methods of preventive maintenance structural examination should be evaluated and appropriately selected.

12.2.1 Preparation and Planning

A basic understanding of the duct structural system's design and intended behavior can be beneficial when coupled with a preexamination site visit. Every detail of an examination should be fully developed well in advance so that valuable plant outage time is not sacrificed for lack of access, equipment, and support personnel. Allowances for cool-down of the ductwork, owner checkout, and safety training should be factored into the schedule. Field preparation activities, such as the installation of temporary lighting, ladders and scaffolding (or equipment rental, such as boom lifts like a JLG or scissor lifts, where scaffolding access is not cost-effective), and the cleaning of interior surfaces by vacuuming, sand-blasting, water-washing, or grinding, should be planned ahead of time to reduce the critical outage time committed to these tasks. Also, if the examination is planned well before the outage, the amount of ash or sludge accumulation can be documented. The availability of laborers to assist with the examination efforts should also be established in advance with plant personnel, along with the schedule of any other work that is planned inside the duct or in the adjacent areas, to ensure adequate coordination and avoid hampering any work being done concurrent with the examination.

Preparations should always address personnel safety. Planning for safety considerations (described in the next section) is essential and should be coordinated with the owner.

Daily or frequent meetings between the plant personnel and the lead structural examining engineer are essential to enable clear communication throughout the examination. Final planning should also include the preparation of contingency plans in the event that the scope of an examination must be changed during the effort.

12.2.2 Safety Considerations

The procedures for duct examination should address the owner's responsibility and safety requirements, plant-specific safety training and safety procedures, and current US Occupational Safety and Health Administration (OSHA) requirements. Three OSHA regulations (part of OSHA Standard 29) pertain to duct examinations: *Permit-Required Confined Spaces* (CFR 1910.146) addresses working in a confined space; *Hazard Communication* (CFR 1910.1200) addresses exposure to hazardous materials; and *Personal Fall Protection Systems* (CFR 1910.140) establishes performance, care, and use criteria for personal fall protection systems. Tests for the levels of hazardous materials should be performed with, or by, the owner. Preferably, body-worn oxygen meters are used while performing examinations inside ductwork to provide continuous monitoring of the air quality. If this information is unknown and is not to be determined, the worst conditions should be assumed.

Project-related duct examination procedures should also identify other precautions, including working in areas of hot ash; soft deep ash; items weakened by corrosion or erosion, such as handholds, ladders, handrails, and duct floor plate; slippery surfaces; and abrupt changes in duct direction or elevation dropoffs. Also important is understanding the need to wear proper face masks, respirators, fall protection equipment, and all other personnel protective clothing and equipment. Certain hazardous elements in the ash may require the use of full-face or half-face respirators or even self-contained breathing apparatus during the inspection. Preplanning for these items is imperative because medical examinations for physical fitness to use these devices may be required of the inspection team.

12.2.3 Equipment

The selection of equipment for a preventive maintenance examination is subject to the project scope and available resources. Table 12-1 lists some of the equipment commonly used for effective and efficient preventive maintenance examination of ductwork.

Items to facilitate access, such as ladders, bosun chairs, scaffolding, safety harnesses and lanyards, and climbing cages, should be available as appropriate. These items are normally provided by the plant. Planning ahead to arrange for the use of these access items is just as important as providing the items on the suggested equipment list (Table 12-1). The use of plant labor to assist with the examination effort is also important and must be well coordinated with the plant personnel.

Table 12-1. Structural Maintenance Examination Equipment Checklist.

Flashlights (hand and helmet)	—	Ultrasonic thickness detector	—
Flashlight holsters	—	Liquid dye penetrant test kits	—
Camera with flash (or tablet)	—	Infrared imaging equipment	—
Cell phone	—	Inclinometers	—
Video camera (or tablet)	—	Thermocouples	—
Steel-toe work shoes	—	Accelerometers	—
Binoculars	—	Spark testers	—
Tape measures (manual and laser)	—	pH meter	—
Calipers	—	Weld profile gauge	—
Thickness gauges	—	Ladders	
Chipping hammers	—	Bosun chair	—
Picks and shovels	—	Scaffolding	—
Wire brushes	—	Safety harness and lanyard	—
Stationary mirrors	—	Lifeline	—
Telescopic mirrors	—	Climbing cages	—
Plumb bobs	—	Eye protection	—
Clipboards	—	Dust masks	—
Pens (many)	—	Gloves	—
Ruled paper	—	Coveralls / disposable cloth suit	—
Cloth rags	—	Keychains	—
Crayons, chalk, spray paint	—	Two-way radios	—
Markers	—	Hard hats	—

Installing thermal displacement measuring devices at strategic locations on the ducts is often helpful. These devices are commonly called trams. Obtaining cold and hot duct location measurements from a fixed reference point will indicate the amount and direction of thermal expansion. Comparing the actual thermal movements against the theoretical design thermal movements can help identify the root cause of a duct (or a duct apparatus such as a damper) not functioning as designed.

More specialized equipment may be warranted for more comprehensive or unusual examinations. Temperature-indicating crayons may also be of value. For the examination of duct linings, spark testers, pH meters, and thickness gauges for thin material measurement are typically used.

12.2.4 Identification of Critical Duct Sections

Transition sections, elbow sections that abruptly change directions, and manifolds are examples of duct sections that may require a more rigorous examination. Depending on the duct system's design and configuration, certain sections may experience greater-than-expected loadings, flow velocities, expected temperatures, stresses, and thermal movements. With continued operation, these sections may not perform as the structural design engineer intended, and they may undergo significant wear, deformation, and deterioration.

The type and arrangement of supports are important considerations in determining the scope of a duct maintenance examination. Duct systems may be hung by variable-support or constant-support hangers, or they may be bottom supported. Sometimes ducts are supported by a combination of these. The degree of structural determinacy is often related to the potential for structural distress. Ducts with minimal supports (four to six) are more likely to behave in a manner consistent with the structural analysis. The behavior of indeterminate ducts with many supports tends to be quite different from that predicted by classical structural analysis. Such differences are the result of thermal growth or restraint, relative stiffness of the support structure at the duct support points, secondary stresses, and the stress induced by the construction sequence. The arrangement of the structural supports, the means by which the system loads are distributed to the supporting structure, and the degree of redundancy all contribute to a duct section's structural behavior. These items may result in actual applied loads exceeding the original design, so additional attention should be given during examination to ensuring that an adequate load path exists and that there is no indication of overstressed members.

Other critical duct sections include areas with internal flow distribution devices. Concentrated erosion may occur at these sites, reducing the thickness of structural components and threatening their structural integrity. In some cases, the flow distribution devices also serve as internal struts for structural load transfer or wall panel support. The location of these devices within the duct system and the amount of particulate in the flow are factors to consider when prioritizing maintenance examination tasks.

Any preventive maintenance examination should include the points of entry into the ductwork, including access doors, grab bars, guardrails, and platforms. The examination should also include all ductwork probe penetrations. Because the condition of hatches and access doors is critical to both personnel safety and the system's gas-tightness, the examination should include all seals. Ductwork probe penetrations for items such as thermocouples and gas test sample ports offer possible areas for accelerated deterioration from air or water intrusion or cracking from high stresses or vibrations.

12.2.5 Exterior Walkdowns

Observation of the external features of a duct system can provide valuable information on the duct's behavior and condition. This type of examination may be performed in advance of a plant outage and the subsequent detailed internal examination. In many cases, indicators of potential damage can be noted on exterior walkdowns, as discussed in Section 12.3.1, which may identify problems or locations that should be emphasized during a more comprehensive or detailed exterior and interior examination. Removal of selected insulation and lagging may be necessary to perform an initial walkdown. This is usually only done after the unit comes out of service. The duct examiner should also be cautious before coming into contact with any ductwork component while the unit is operating, because it may be very hot. This is particularly true at duct support legs, ports, and other protrusions where additional heat loss is likely to occur, but may also apply to the duct lagging if the insulation is deteriorating or the insulating system was inadequately designed.

Exterior walkdowns should also include a cursory examination of duct system elements that are remote or inaccessible from grade or viewing platforms. Binoculars or a video camera may provide adequate viewing for these areas. A cursory examination may substantiate the need for a comprehensive examination, calling for the erection of scaffolding, climbing cages, or ladders. If the structural engineer performing the examination believes that a comprehensive examination of certain remote areas is required, the required access should be provided within reason.

When possible, both a hot (on-line) and a cold (off-line) exterior walkdown should be considered to observe and record the movement and thermal growth during load and temperature variations. Baseline measurements while the plant is off-line, referenced to permanent stationary structures, may provide valuable insight on thermal growth and support alignment. Displacement measuring trams are normally used to collect these readings.

12.2.6 Interior Walkdowns

Assuming adequate access, internal structural elements of the ductwork should be examined in a detailed manner. The condition of critical structural elements (described in Section 12.3.2) should be determined and accurately recorded. A more detailed condition assessment of certain duct internal structural elements can be performed on the basis of documented observations. Other than the obvious condition of structural elements, observations worthy of documentation include the distribution and depth of ash accumulations, surface discolorations, and the general alignment of parallel members and walls.

Careful planning is required for a comprehensive examination. Because the internal examination may be limited in time and schedule to coincide with a unit outage, the examination team should plan their work in advance to work as quickly and efficiently as possible. This includes setting up field notes before the interior walkdown to minimize the documentation and recording time. It may also be necessary to make arrangements with the plant to perform the internal walkdown around other activities occurring on site, whether inside or outside the duct.

12.2.7 Thickness Measurements

Both internal and external duct elements may experience corrosion and erosion. Field measurement of the material thickness is not always practical, but thickness measurements are essential for determining an element's current structural properties to evaluate its load-carrying capability. Local loss of thickness in steel shapes and plate can result in weak elements and offers locations for future failures. All paint, corrosive matter, ash or sludge, and general debris should be carefully removed, without damaging the element, to obtain accurate thickness readings. If extensive cleaning and preparation will be required, detailed planning should be performed to select, gain access to, and prepare these areas for measurement. Some areas may require the drilling of holes to allow thickness measurements to be taken with calipers if an ultrasonic thickness detector is not available or practical to use. Plans should include capping or plugging these test holes after completion of measurements.

12.2.8 Temperature Monitoring

The temperature of flue-gas or air in ducts can vary considerably from the established design basis. In some instances, the expected temperatures and the actual on-line temperatures may vary enough to warrant field measurement. Also, unanticipated thermal stratification and temperature differentials may occur within the ducts. Specific localized temperatures of elements in the duct can be obtained and recorded instantaneously over a period of time through the installation of thermocouples. Thermal records developed during plant start-up and shutdown, as well as those associated with steady-state operations, often identify thermal gradients. During start-up and shutdown, structural elements may experience a temperature lag when compared to adjacent elements in the system. This can also be identified if the data are collected properly. When strategically placed and properly installed, thermocouples may provide information that explains the structural behavior of the duct.

During external on-line examinations of operating ducts, temperature-indicating crayons or digital thermometers can provide accurate measurements of hot and cold spots on the lagging and duct plate.

12.2.9 Displacement Measuring

To collect information on the actual thermal growth and expansion of the ducts, thermal displacement measuring devices, or trams, should be installed at strategic locations on the ducts. The best locations are usually at the corners or ends of each duct section. The trams must be installed such that they are referenced to permanent stationary structures. They should be installed as far from each other as practical to measure significant movements. Readings should be obtained when the ducts are at their operating temperature and when the plant is off-line. Comparing the cold and hot duct location measurements will indicate the amount and direction of thermal expansion.

12.2.10 Infrared Survey

Infrared surveying of the exterior of the ducts can be performed during an exterior walkdown while the plant is in operation. This survey can provide information on air or gas leakage and poorly installed or deteriorated insulation and lagging. Qualified operators of this equipment can detect local hot spots, which can generally be correlated with a structural concern or developing structural problems.

12.2.11 Laboratory Test Samples

In some instances, samples of duct steel should be obtained for chemical and physical laboratory analysis. A means to obtain, identify, and transport the samples should be a part of the comprehensive duct examination plan. The desired laboratory tests should be specified well in advance of the examination. These tests include those associated with material verification or classification, strength, creep, impact properties, and contamination.

The results of the material testing program should be used as part of any subsequent assessment of the duct system's structural integrity. Comparison of the results with the established design basis, industry standards, and available historical or academic data may offer insight on the material strength, serviceability, and rate of deterioration.

12.2.12 Documentation

The investment in preventive maintenance ductwork system examinations can be enhanced by the complete and accurate recording of the collected information. First, a procedure should be created to properly record the examination field notes. This procedure should address basic considerations such as the uniform use of pens, paper, and markers, and the type, format, and desired detail of notes and abbreviations. Specific instructions should include the exact type of information to be recorded, such as references, orientation, sizes, type of damage, and visual observations.

The presentation of information on previously developed data sheets will facilitate collection, compilation, and evaluation of the examination field notes. Where practical, markers, chalk, or paint may be used directly in the duct system and support structure to highlight critical findings or merely indicate that a portion of the examination is complete.

Photographs and video recording of damage and any unusual observations are highly recommended. Good and safe lighting should be provided to enable high-quality photographs.

All information collected during a duct field structural examination, including photographs, should be included in the final report. See Section 12.4.4 for more recommendations regarding the final report. This report should serve as an important reference document for the continued future evaluation of the ductwork system.

12.3 POTENTIAL DAMAGE AREAS

The structural condition assessment of ductwork should include an investigation of any indicators of potential damage areas, as listed in Section 12.3.1. The presence of some of these indicators may warrant a detailed, rather than a cursory, examination of the key structural elements in the area of the indicators. See Section 12.3.2 for discussion of the examination details of these key structural elements.

12.3.1 Indicators

The following items should be examined during a structural maintenance examination, as they might be indicators of adjacent structural damage.

Lagging. Visual observation of the lagging can provide information on the condition of a duct structural system. Indicators of potential structural problems include localized stains, general deterioration, buckling, separation of seams, binding at supports or slide plates, and the ponding of water. The topographic condition of lagging surfaces can be an indicator of the configuration or settlement of the duct surface beneath the lagging. Sometimes lagging can be missing altogether, which creates obvious points of moisture intrusion.

Insulation. Some duct systems have only insulation and no lagging. Again, a preventive maintenance examination should identify areas with missing, damaged, or nonuniform insulation. Hot spots that cause stress variations in these duct sections, and cold spots that accelerate corrosive deterioration, may then be avoided or eliminated by repairing the insulation.

Expansion joints. Cracked or torn expansion joints may indicate erosion, ash packed and hardened in the joints, or unusual duct movements. A more careful examination of the duct supports adjacent to damaged joints may be warranted to determine the extent of any duct damage and the size of duct support thermal movements. The material and configuration of any damaged expansion joint should be investigated to assess its compatibility with the operating environment. Ash or sludge buildup in expansion joints is a frequent cause of damage. Once the joints get packed with ash or sludge, the duct sections may be unable to freely expand, and thus may become damaged. Erosion shields or covers for thin metal joints, and liners or baffles for nonmetallic joints, should be examined for deterioration and ash buildup.

Expansion joint frames should also be examined for damage. Damage to these rigid frames is usually caused by improper insulation details, improper joint seals, or a packed joint. See Chapter 11 for discussion of proper insulation around expansion joints.

Condition of coatings and linings. Small cracks in surface treatments can indicate the direction of stresses in an element. A detailed examination report should include an accurate depiction of such patterns. Peeled paint on adjacent support steel could indicate that the steel has been subjected to very high temperatures, and the spalling of paint may indicate the buckling of underlying plate. If the support steel is painted in the vicinity of sliding supports, worn paint surfaces can accurately indicate the amount of thermal growth.

Excessive deflections. Duct structural internal members may exhibit vertical or horizontal buckling, deflection, or rotation. External stiffeners may become buckled, warped, or severely bowed. Member web crippling and excessive connection deformation may also be observed. In all of these cases, the condition of the member and its associated end connections should be recorded. All sags and bows in the duct plate or stiffeners should be accurately measured during the examination. Accurate measurements of deformations and rotations may help identify potential causes.

Buckling or out-of-plane distortion of duct plates often indicates that the external stiffeners have separated from the plate. If the stiffeners have not separated from the duct plate, they are usually quite damaged. Distorted roof plates could indicate a loss of lagging integrity. This damaged plate is usually caused by ponding of external water, or thermal gradients associated with poorly distributed flows.

Fly ash and scrubber sludge accumulation. Particular attention should be given to areas where high volumes of fly ash and scrubber sludge accumulate. These areas should be examined before and after removal of the accumulations. Because the sludge acts as an insulator, duct floors may be cooler than the wall and roof sections. Temperature differentials and corrosion of the duct plate can occur in these locations. Duct corners should be closely examined for cracks in the duct plate, welds, stiffeners, and corner angles, because large thermal stresses build up and produce the greatest damage in areas lacking flexibility.

The amount and locations of significant deposits should be recorded. Samples may be collected if density tests are needed to accurately determine the weight of the accumulation.

Discoloration. Interior duct surfaces frequently exhibit a reddish-brown or gray color. The examiner should note blackened areas, which may indicate a prior overheated condition such as a fire. White or very light-colored areas indicate severe inward leakage of water. In both of these areas, the material properties and the strength of adjacent elements may have been affected. Structural damage might also be present.

Corrosion. Internal duct elements may be subject to extreme corrosion. Corrosion may be attributed to operating temperatures or local plate temperatures below the dew point of the flue-gas. Because corrosion will reduce the strength of the plate and elements, the safety of the examiners should be considered throughout an examination, especially in FGD system ducts.

External duct members may be subjected to corrosion from external water sources.

Erosion. Internal duct members are subject to abrasion from particulate matter carried by the flow. The conservative design of these members, as well as the attachment of sacrificial shielding members, can provide additional protection and prolong service life. However, almost all internal surfaces are subject to some degree of erosion. Noticeably worn or polished areas typically develop at flow distribution devices, in corners, and in areas where the duct configuration changes. Local areas of high velocity and high particulate content should be examined to ascertain the extent of any erosion.

Spalled or honeycombed mortar or refractory. There are different concerns for ducts with internal mortar or refractory insulation. Mechanical attachments are used to anchor the insulation to the steel plate. Damage to the duct plate usually starts at the location of the attachments. Worn areas, sags, and areas believed to have separated from the duct plate should be documented. These areas can be identified by rapping the interior surface with a hammer. Flue-gas leakage to the protected duct plate frequently occurs at areas of cracking and spalling. These are sites for condensation and inevitable corrosion, which can be correlated with exterior duct plate deterioration. Spalling or honeycombed mortar also indicates that the refractory may be locally unstable. Unusual deterioration may warrant taking refractory core samples for strength testing and petrographic analysis.

12.3.2 Structural Elements

The following structural elements should be examined during a structural maintenance examination.

Supports, slide bearing plates and guides. All components of the duct's support system should be carefully examined. The condition of the various types of bottom supports may reflect different types of operational concerns and history of events. For example, slide bearing plates with scorch marks and peeling surfaces would indicate that very high temperatures have existed at this location during operation. Guide bars at sliding supports should be examined for cracks in welds or plates, permanent deformation, and wear marks. Where visible, low-friction surfaces should also be examined for large scratches, gouges, delamination, acid holes, and bulges at the center of the plates. When examining off-line duct systems, thermal movements can be determined by measuring the clean areas of slide plates that develop as the ducts cool and contract from the operating position. Comparison of these measurements with those theoretically predicted is an essential method of assessing the duct's structural behavior.

Hangers. Both variable-load and constant-load hangers are used to support ducts. Loads other than expected, excessive load changes, or excessive vertical movement may occur in hangers that could result in damage to the hanger, the hanger support steel, or the duct at or adjacent to the hanger attachment.

The condition of the hanger connections to the duct and to the support steel should be examined. The hanger manufacturer, model number, and spring constant should be recorded. For each hanger, the vertical movement and the position of the indicator pin should be recorded for both cold and hot conditions.

Stiffeners. When accessible, duct stiffeners should be visually examined for continuity, deformations, and corrosion. The weld attachment to the duct plate should also be examined to ensure there is a positive connection. Sighting along the length of members will identify any bows. Close examination is required to find any cracks. For critical members, the extent of the bow can be determined by tensioning a string along the member's length and measuring the offsets. The end connections should also be examined for cracks, tears, and deformation.

Internal bracing. Bracing members may be comprised of double angles, wide flange members, structural pipes, or structural tubes. An examination should identify and measure the extent of bows as well as any deviations from the design drawings. Each member's connection plates should be examined for cracking, buckling, stability, and sound attachment. Welded connections are typically used for these members, and all welds should be carefully examined for cracks.

Structural bracing members should also be checked for thin sections caused by erosion. In some instances, sacrificial wear plates or angles may have significantly deteriorated, jeopardizing the parent member's structural integrity. Understanding the original design criteria and approach to these members is important, because in some cases the design may have included sacrificial erosion/corrosion allowances. Examiners should be conscious of how a member's orientation and its associated connections restrict the flow.

Plates and welds. All accessible areas should be examined for cracks and general deterioration. Plate cracks generally occur adjacent to welds in the heat-affected zone, at corners, and near the duct supports. These are areas where the structural system has greater rigidity, and residual stresses may be present. The extent and orientation of surface cracks may be determined by the use of liquid dye penetrant or magnetic particle methods. As mentioned in Section 12.3.1, the condition of painted surfaces may also indicate underlying plate cracking.

Bolts. Bolted connections should be checked for looseness, distress, or failure. Each bolt should be checked for shear or tension failure or any observed deformation, engagement, looseness, or deterioration. The amount of bearing or contact area should also be examined. In sliding connections, bolts should be plumb, but loose enough to enable the design movement to occur. Missing or deformed bolts may indicate support overloads or irregular holes. To check whether bolts are loose, each assembly can be twisted or rapped with a hammer. The observation of bolt classification markings may be useful where different strength bolts may have been used.

12.4 EXAMINATION DATA, EVALUATION, AND DISPOSITION

Field data collected in the examination should be neat, complete, accurate, carefully compiled, carefully evaluated, properly dispositioned, and organized into a final report for presentation to the owner. A set of ductwork drawings can be helpful in documenting field observations and photo locations. The field notes should also include maps of the noted damage so that it can be easily located later by the evaluation engineer, the owner's representatives, and the repair contractor.

12.4.1 Examination Data Assimilation

Field data should be completely and accurately recorded. Photographs or video should be taken of defective, damaged, or problem areas. Field notes should indicate the location and date of each photograph and the limits of the performed examination. Inaccessible regions should be properly documented and discussed in the final report. Pertinent data, such as a means of picture identification, orientation, reference directions, sizes, lengths, and measurement scales, may be marked on or near the item being photographed prior to photographing.

Repair maps created from accurate sketches or markups of the original drawings, showing dimensions and orientations, should be compiled for all areas examined. The maps of all examined areas will enable the correlation of field data to specific areas of the ducts. Maps showing the locations of thickness measurements are also required. Examination forms should be prepared ahead of time to reduce the time needed to record field notes and sketches.

12.4.2 Repairs

Where damage is found, structural repairs are usually required. These repairs may be designed for a short-term or long-term life, depending on the circumstances. These repairs are necessary to render the duct safe for continued plant operation. Engineering sketches, in lieu of detailed design drawings, are often used to specify repairs quickly and efficiently in the field. Such sketches are usually urgently required to allow the ductwork to be returned to service quickly. The preparation of such sketches should not bypass appropriate quality control procedures of the inspecting company or owner or ignore the proper responsibilities of the engineer of record (such as performing calculations where necessary).

Before an examination, generic or standard repairs for anticipated damage should be prepared. Such an approach will minimize valuable examination and outage time dedicated to the development of repairs. Standard details will also promote uniformity and an orderly and cost-effective construction process.

Some repairs can be implemented while the plant is operating. They include the replacement of deteriorated lagging and insulation and the modification of lagging. This is assuming that access and surface temperatures can be handled properly. Depending on the extent and magnitude of the damage and deterioration, outage-based repairs may be prioritized on the basis of criticality over several future outages. The lead times for replacement parts, such as expansion joints or certain structural members, could impact the scheduling of repairs. Any significantly damaged structural members should be addressed by immediate reinforcement or complete replacement.

Given the need to assure the stability of the duct system and the safety of plant personnel, many of the repairs or modifications may need to be coupled with the design and installation of temporary support structures. In the case of refractory-lined ducts, selective repair or removal may be the most effective approach when coupled with the resealing of areas where flue-gas leakage could occur. After repairing corroded areas or areas with deteriorated coatings, cleaning and recoating may be appropriate to prevent future deterioration.

12.4.3 Evaluation and Disposition Considerations

Many factors must be considered regarding the disposition of structural examination findings. Potential issues for resolution include the severity of damage or deterioration, future monitoring plans, the commitment of non-outage versus outage repair efforts, time constraints, capital expenses, and alternatives to plant operation philosophies. The purpose of a thorough condition assessment goes beyond providing an immediate repair or estimating the frequency of future repairs. A full assessment should address the causes of damage, identify any concerns, attribute documented damage to one-time occurrences or ongoing mechanisms, and make recommendations for long-term corrective action.

Severe or unexpected damage will most likely require further evaluation. In some cases, a structural analysis of the duct system might be necessary that incorporates those components with severe corrosion, buckles, cracks, or excessive deflections. Calculations and pertinent industry and academic data may be applied to predict the rate of deterioration. Then the remaining useful service life of duct structural components may be estimated, provided that all the factors that may impact the estimated design life are carefully considered in this determination. In reality it is difficult to predict the remaining life, given unknowns such as projected operating conditions (pressure, temperature, duration, flow rate, etc.), actual material properties (physical and chemical), possibility of abnormal events that may subject the duct to temporary conditions that exceed the original normal design conditions, number of vibratory cycles the material has been exposed to, large variability in the actual creep-rupture test data that design properties are derived from, redundant load paths, and so on. These unknown conditions severely undermine the accuracy of the estimated design life, and therefore an estimated design life is not typically determined.

A comprehensive evaluation should also differentiate between localized and potentially progressive failures. The evaluation should include a comparison of actual service conditions and loads, as observed in the examination, to those assumed in the original design. Stress intensity must be an integral part of the evaluation process for deteriorated members. Their remaining factor of safety, if any, must be determined. Finally, if the documented damage is deemed catastrophic, a detailed assessment should consider alternate modes of unit operation as a means of resolving the potential dangers.

Alternative resolutions to repairs, whether minor or major, often have to be coupled with a change in operating conditions. This option is quite drastic, and therefore is rarely considered viable. However, should extreme conditions warrant this type of evaluation, additional multidisciplinary input should be sought, with special attention to the experience of plant operators. Similarly, certain problems may be of a chronic, recurring nature and may never be resolved without a complete redesign of the duct system. However, experience has proven that most damage and deterioration can be reduced through the development of a systematic, routine and comprehensive preventive maintenance structural examination.

12.4.4 Final Report

The final examination report will be an important future reference and should be comprehensive. All aspects of the preventive maintenance examination effort should be included. All field data, repair sketches, and analysis work, if any, should be summarized and also presented in all its detail. All photographs and video should be included as attachments. All the information should be concise, clearly presented, and carefully arranged. The report should include all conclusions and recommendations for future courses of action. Following is a suggested format for the final report. Given the time and effort that may be required to prepare this important document, exit interviews with plant personnel are frequently conducted as a preview to the final report.

Suggested Final Report Table of Contents

1. Purpose
2. History
3. Summary
4. Scope and Methods of Examination
5. Details of Examination
6. Conclusions and Recommended Future Course

Attachments:

(a) Access Sketches

(b) Damage Report

(c) Inspection Photographs and Video

(d) Repair Sketches or Original Drawings Revised to Show Repairs

References

OSHA (Occupational Safety and Health Administration). 2011. *Permit-required confined spaces.* 29 CFR 1910.146. Washington, DC: US Dept. of Labor.

OSHA. 2013. *Hazard communication.* 29 CFR 1910.1200. Washington, DC: US Dept. of Labor.

OSHA. 2019. *Personal fall protection systems.* 29 CFR 1910.140. Washington, DC: US Dept. of Labor.

CHAPTER 13

Reinforcement of Existing Ductwork

13.1 INTRODUCTION

As flue-gas emission requirements become more stringent, existing boiler exhaust systems may require new environmental control systems, such as precipitators or fabric filters, scrubbers, SCRs, and potentially new ID fans and booster fans. These new environmental controls and additional fans will likely increase operating pressures within the existing boiler exhaust ductwork. Modifications and/or reinforcements to the existing ductwork systems may be required to accommodate these new configurations and new environmental control systems. Similarly, fuel conversion projects at existing coal-fired generation facilities may also require modifications and/or reinforcements.

Portions of the existing ductwork may need to be modified to accommodate an attachment to a new piece of equipment, or a new opening and turning vanes could be added to send flue-gas in a new direction. Other circumstances may allow the entire ductwork configuration to remain unaltered but necessitate reinforcement of key ductwork components (such as duct plate, stiffeners, support stubs, support assemblies, and trusses) to accommodate the new design temperatures or pressures after the addition of environmental control equipment or fans. Whether the ductwork configuration is modified or the ductwork components reinforced, careful evaluation and structural assessment are necessary to ensure a sound structural system in the final design condition and configuration. It may be necessary to carry this evaluation over to the existing ductwork support structures to assure that they are adequate for the changed loading conditions of the reinforced or modified ductwork.

Sometimes, reinforcement/retrofit of existing ductwork components may save on material costs and other indirect construction costs compared to wholesale replacement. In other cases, complete replacement of whole ductwork sections may be selected to simplify erection and reduce the outage time. An evaluation should be performed to ensure that modification is more cost-effective than a complete demolition and replacement of the ductwork. In general, all relevant project team members, including the engineer, owner, and installation erector, need to be involved in deciding whether to modify/reinforce or replace.

13.2 EVALUATION OF EXISTING DUCTWORK

An evaluation of the existing ductwork may identify structural deficiencies and inadequate ductwork components that will need to be addressed given the new design conditions. The process is similar to that used for design of new ductwork; but the ductwork components are known and are evaluated under the new design conditions (temperature, pressure, flow velocities, configuration, etc.) and current codes. The evaluation process begins with a determination of the new design loads and appropriate load combinations, followed by global analysis of the existing structural system, and finally local analysis of individual ductwork components for the new anticipated loading, boundary conditions, and structural behavior. All existing ductwork and supporting structures should be verified to have sufficient structural capacity under the new design conditions. The International Existing Building Code (2017) may be used to determine permitted reductions in the safety factor of the design.

Previous chapters discussed important considerations for each step of the evaluation. Service conditions and design loads are discussed in Chapter 4. Load combinations are discussed in Chapter 5. Ductwork global structural analysis and design is discussed in Chapter 6. Finally, local structural element design and strength evaluation is discussed in Chapters 7 and 8. Inspection of the existing ductwork is discussed in Chapter 12.

13.3 DUCTWORK WALKDOWN

A walkdown should be performed of the ductwork that is to remain under the new design conditions to accurately determine the as-found condition of the ductwork and to ascertain any missing information not explicitly provided on the design drawings. The ductwork configuration and components shown on the design drawings may not always match the as-found condition in the field. A walkdown will help identify undocumented design changes, modifications, and potential fabrication errors, each of which could significantly impact the structural analysis and evaluation of the existing ductwork.

As the ductwork may have been in service for many years prior to the evaluation, individual ductwork components may have experienced some corrosion or erosion over time. Depending on client preferences and the as-found condition of the ductwork, it may be prudent to include material loss caused by corrosion or erosion in the evaluation of the ductwork.

For ductwork with indeterminate supports, it may also be necessary to consider loss of support bearing, if the walkdown reveals evidence of arching of some duct supports in either operating or nonoperating conditions. A walkdown may identify alternate load paths that are not consistent with the original design assumptions and that may need to be addressed in the evaluation of indeterminate ducts.

More information on walkdown planning, safety considerations, equipment, and documentation is given in Chapter 12.

13.4 DUCTWORK MODIFICATION

Sometimes an existing section can be modified for the new ductwork configuration. For example, new ductwork may tie into existing ductwork between the economizer and airheater to redirect the flue-gas flow to a new SCR. The modification may include installation of blank-off plates, cutting of new openings, and installation of new trusses and turning vanes, among other modifications. Rather than demolishing all of the duct section, it may be more cost-effective to modify the existing ductwork during a brief outage.

Most modifications to existing ductwork should be performed during a unit outage, as opposed to during operation. Welding of reinforcements to operating ductwork involves two major areas of concern for the engineer/contractor: the safety of personnel performing the reinforcement work and the unpredictable performance considerations, potentially effecting the structural integrity of the ductwork. If welding of reinforcements is to be performed while the unit is in operation, the safety of the personnel should be carefully reviewed and monitored by the responsible contractor and the requirements of the applicable AWS codes and OSHA regulations should be implemented. Moreover, from the engineering perspective, there are several technical challenges that are difficult to resolve with confidence when welding two elements at significantly different temperatures. Among the issues that cannot be accurately analyzed or predicted are the following.

- Locked-in stresses may result from welding relatively cool unstressed elements to operating ductwork. These stresses stem from both the thermal gradients and the in situ stresses (including that of system pressurization) and affect the duct plate, stiffeners, and connections. They can promote premature weld and plate cracking, flue-gas leakage, duct corrosion, and decreasing structural strength of ductwork. Although the general structural behavior of welding to operating ducts is understood in the heat-affected zone of the weld, the magnitude of overall member thermal gradients, along with resulting global locked-in stresses and their negative impact on duct strength, cannot be accurately determined.
- Removing insulation in local areas will increase thermal gradients in the operating duct, potentially creating even higher secondary stresses and structural distortion, and structurally weakening the ductwork.

- Although these additional stresses may result in overall member stresses within the yield strength of the material, the effect of multiple start-up and shutdown cycles, combined with the locked-in stresses, could lead to damage at an unknown point in the future.

Given the difficulties and uncertainties associated with analyzing and designing reinforcements in the unpredictable environment that occurs when welding to operating ductwork, it is recommended that welding of ductwork reinforcements be performed while the ductwork system is off-line.

In certain situations, the flue-gas path may need to be reconfigured and the flow turned in a new direction. The unbalanced pressure forces at the ductwork bend can drastically change from the original condition and require new load paths from the ductwork to the structural steel and foundation. In some cases, new trusswork may be required at duct plate openings to continue the existing load path along the duct wall to existing duct supports. In other cases, duct supports can be relocated or added to reduce reinforcing to take the new unbalanced pressure loads. The new duct supports shown in Figure 13-1 take the unbalanced pressure force to reduce the amount of reinforcement required or to prevent new loads from being applied to the existing support steel or foundations. In adding a new support, the thermal movements of the duct must be carefully considered, to ensure that adding the support will not have adverse effects on the behavior. Every project is unique, and no one modification, replacement, or reinforcement approach will fit all circumstances. The structural engineer should work closely with the owner and installation contractor (if applicable) to determine the needs of the project and consider cost, schedule, and constructability.

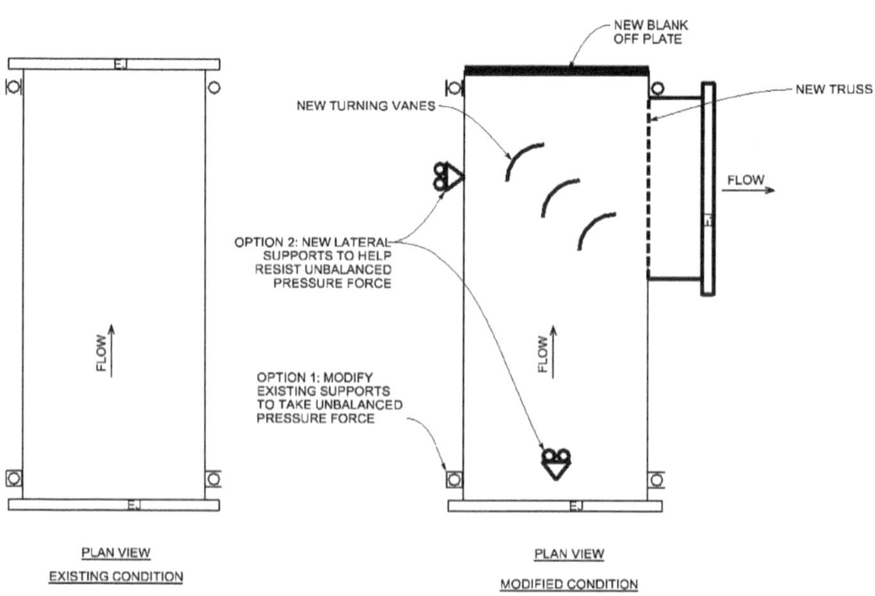

Figure 13-1. Example ductwork modification from straight run to elbow.

13.5 REINFORCEMENT

13.5.1 General

Selection of reinforcement methods and details varies greatly, depending on project specifics. Existing ductwork and support steel layouts may restrict or even prevent certain reinforcement methods. Each situation is unique and should be considered individually. Typically, reinforcement is installed during a unit outage, so the ease and speed with which that can be done is very important. Internal modifications may be preferred to reduce the amount of insulation and lagging that needs to be removed and reinstalled and might reduce the unit outage time and construction costs.

Reinforcement is required for all ductwork components that cannot be verified to have sufficient structural capacity under the new design conditions. The following subsections describe common reinforcement practices used to strengthen common ductwork components. The information is not intended to be exhaustive but to provide a starting point for consideration.

When designing reinforcement of existing statically loaded elements, it is important to account for the existing stress levels, also referred to as locked-in stresses. The existing stresses will not be imparted to the new reinforcing element unless the members being reinforced are first unloaded.

13.5.2 Duct Plate

The most common approach to reinforcing inadequate duct plate is to add more stiffeners either parallel or perpendicular to the existing duct stiffeners (Figure 13-2). The new stiffeners reduce the duct plate span and increase its strength capacity.

In some cases, the theoretical duct plate thickness is sufficient for the new design conditions; but isolated thinning of the duct plate, gaping holes, or cracks (typically in hot-side ducts) may have developed over time and should be repaired. This can be done by installing patches or by replacing the existing duct plate. Careful consideration should be given to large patches, in particular in ducts with negative pressures, to ensure that the patch is strong enough to transfer any loads to the existing duct. Additional intermediate stiffeners may be required for strength, given the additional weight of the patch plates. Where isolated cracks have developed, the repair detail should prevent any further propagation of the crack. This can be done by drilling holes at each end of the crack to reduce stress concentrations. Cracks can be filled with weld or patched as necessary.

13.5.3 Duct Stiffeners

Cover plate. A common method for reinforcing duct stiffeners is to install flange cover plates. Depending on the critical loading condition, the cover plate is installed on the inboard or outboard flange. For the inboard flange, the cover plate can be installed on the interior face of the duct plate (Figure 13-3a)—special attention is required to ensure that the new cover plate lines up properly with the

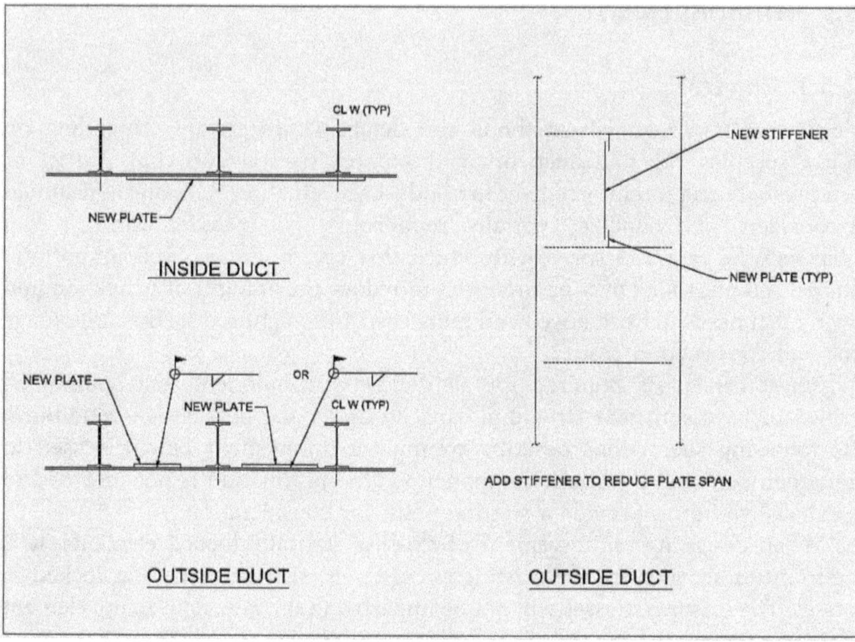

Figure 13-2. Example duct plate reinforcement.

stiffener—or two separate plates on the inboard flange from the exterior of the duct (Figure 13-3b). It may be more economical to use a single larger outboard flange cover plate (Figure 13-3c) instead of the two separate inboard plates, to reduce field welding.

Strongback stiffener. Another way to reinforce ductwork stiffeners is to introduce a "strongback" stiffener (Figure 13-4) to reduce the span of the existing

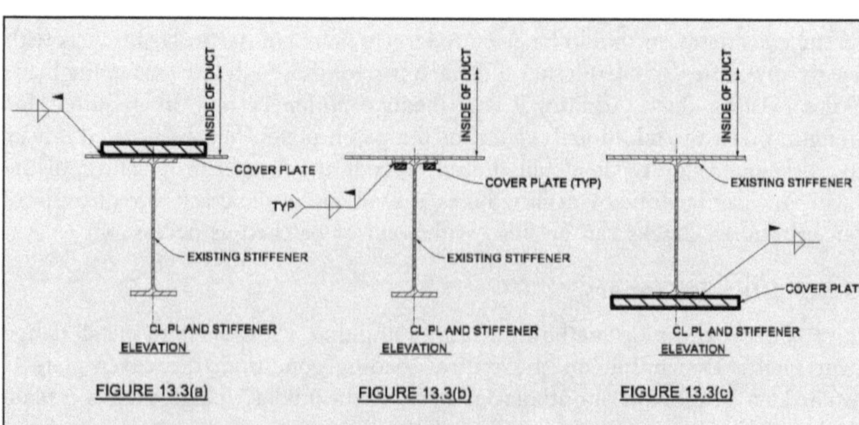

Figure 13-3. Example stiffener reinforcements with cover plates.

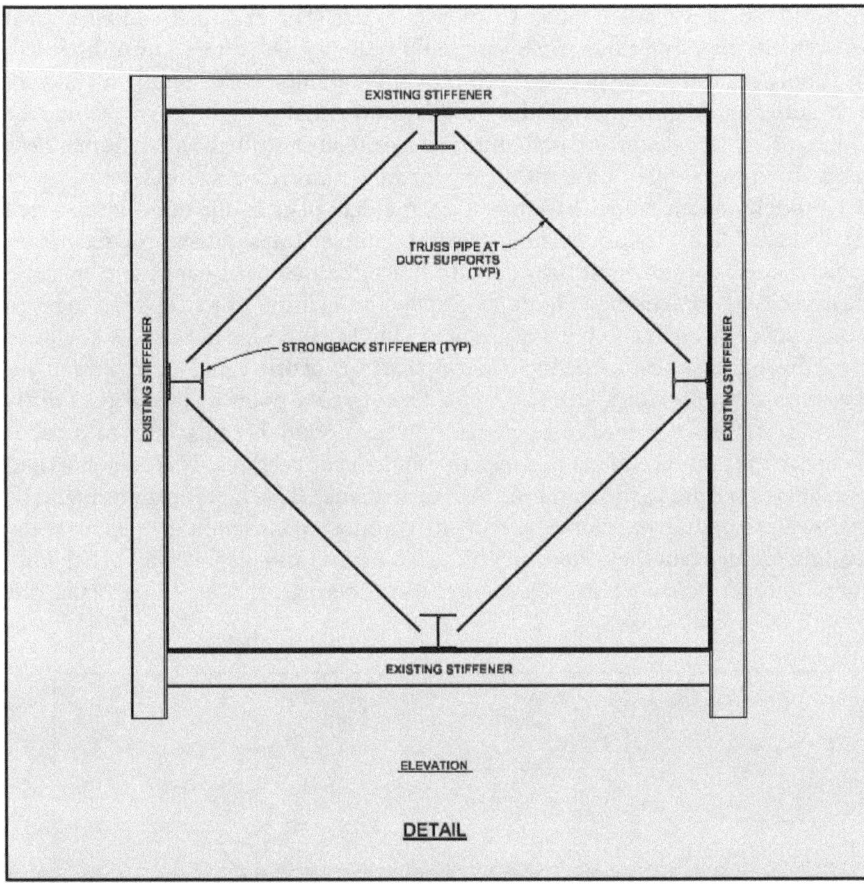

Figure 13-4. Example stiffener reinforcement with new strongback stiffener.

stiffeners. It is installed perpendicular to the existing stiffeners and is typically internally supported with trusswork to transfer load out to the duct supports or to act as a strut as discussed in Chapter 2 to cancel pressure loads acting on opposite sides of the duct. Typically, the strongback stiffener is installed on the inside of the duct to avoid cutting and making modifications to the existing external stiffeners or modifying the insulation and lagging. Load transfer between the existing stiffeners and the internal strongback occurs by bearing against the duct plate, so the strongback stiffener is only effective for negative pressure loads for a typical duct stiffener layout with pinned ends. If an internal strongback stiffener is desired to provide additional strength for outwardly applied loads (positive pressure), a positive connection will be required between the strongback and the external stiffeners.

Struts. Similar to the function of a strongback stiffener, struts can be installed at the midspan of existing stiffeners to cancel pressure loads applied to the two opposite faces of a duct. This also allows stiffeners on opposite faces of a duct to share in resisting loads applied to only one face. For example, adding a strut between the roof and floor stiffeners would permit each stiffener to transfer ash loads only applied to the floor to both the roof and floor, based on the stiffness of each stiffener. Disturbance of the flow and potential erosion issues should be addressed in the design of both internal struts and strongback reinforcement methods. An example of a strut arrangement is shown in Figure 2-20.

Brace outboard flange. In some cases, the duct plate is not sufficient to brace the outboard flange of an existing stiffener. Simple straps, gussets, or secondary perpendicular members can be added to brace the outboard flange and increase the capacity of the member. Figure 13-5 shows an example of a strap used to brace the outboard flange. The bracing element should be designed to have adequate strength and stiffness, as defined in Appendix 6 of the AISC *Specification for Structural Steel Buildings* (2017a), to be considered a point of lateral restraint.

Moment versus pinned connections. Another option that may be considered is to change the pinned end connections to moment connections. This creates a rigid frame to reduce the bending moments in the stiffener. This modification should be investigated carefully, because changing to a moment connection may increase the bending moments in the connecting (i.e., supporting) stiffener. The higher rigidity from a moment connection may also result in thermal stresses (in particular for

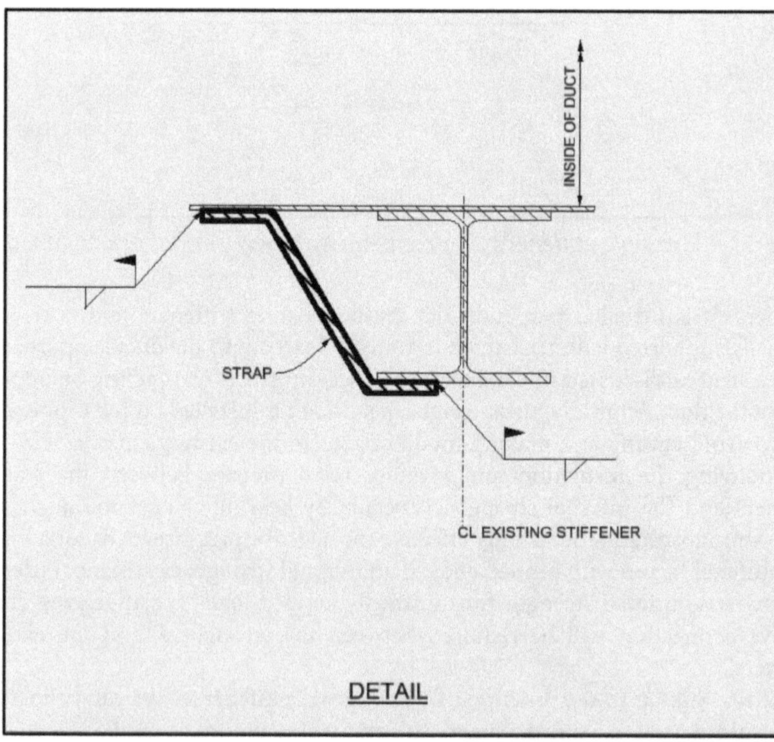

Figure 13-5. Example stiffener reinforcement by bracing outboard flange.

hot-side ducts) that were not present when the stiffener-to-stiffener connection was a flexible pinned connection.

Similarly, a moment connection can be changed to a pinned connection if it is observed that the overall rigidity of the existing stiffener ring is inadvertently developing excessive moments. This typically occurs in the shorter span stiffener in the duct cross section. When an existing moment connection is changed to a pinned connection it should be confirmed that the overall duct stability is still sufficient and that the existing moment connection is not required for the overall global load path system.

13.5.4 Internal Trusses

Sometimes new design pressures or configurations increase the axial strength demand of the existing truss members. Additional cross-sectional area may be required. Figure 13-6 shows examples of cradle, angle, and channel reinforcements for an existing truss pipe. Reinforcements are typically applied to two sides to avoid clashes with the existing gusset plate; but if required, the reinforcement could be slotted around the gusset plate. In some cases, the entire truss pipe may need to be replaced, especially if excessive erosion is observed. The engineer could also consider adding bracing members to reduce the unbraced length of the existing truss pipe, thus increasing the design strength.

13.5.5 Support Stubs

If the ductwork support stubs need to be reinforced for the new design condition and forces, the most common approaches include the installation of side plates or cover plates (Figure 13-7). Reinforcement plates can increase the axial and flexural strength of the support stub, and adding side plates creates a closed shape that significantly increases the torsional strength to resist any local twisting of the support stub that might be caused by the load path in the support assembly or thermal growth of the duct. Typically, the side plates are extended above the bottom of the duct to ensure proper development length for the reinforcement.

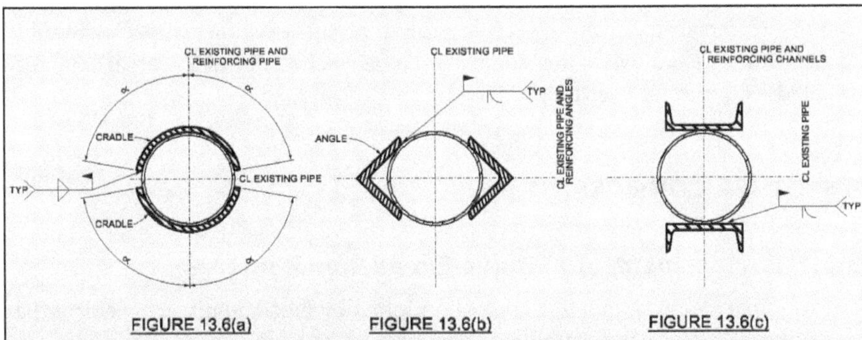

Figure 13-6. Example truss pipe reinforcements.

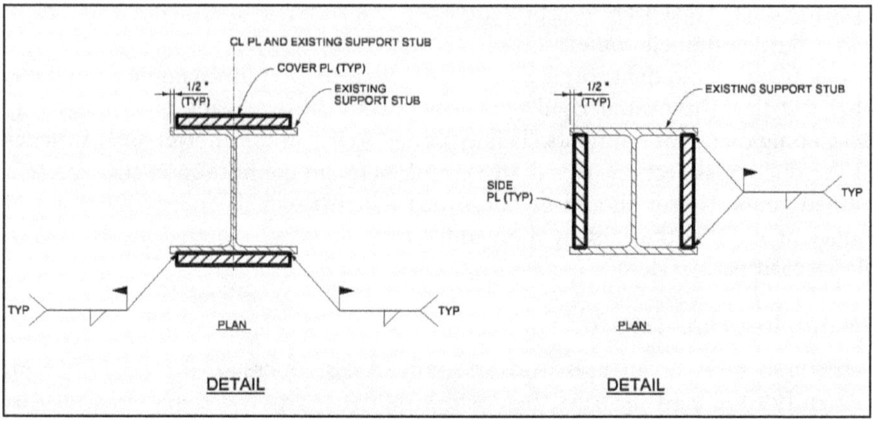

Figure 13-7. Example support stub reinforcement with cover plates and side plates.

Support assemblies (e.g., guide bars) may limit where reinforcement can be installed.

13.5.6 Support Assemblies

Support assemblies should also be reassessed for new loading conditions. Depending on the new loading and the existing condition, additional stiffener plates may be needed for the base plate or guide bars, or larger guide bars may be needed. If the new loading introduces uplift forces where previously uplift was not a concern, additional reinforcement will be required, such as new inverted-L hold-down bars that extend over the existing base plate; or hold-down bolts could be added, with slotted holes to allow for thermal movement; or in some cases new prefabricated uplift restraint slide bearings may need to be considered. The type of reinforcements will depend on the specific configuration of the existing ductwork support, slide plate assembly, and support steel. Figure 13-8 shows an example of new inverted-L hold-down bars welded to the flange of the existing support steel. Any attempt to replace the existing slide plates or the entire support assembly will likely include jacking and temporary support of the ductwork while the support assembly is replaced. Complete replacement of the existing support assembly should be discussed with the client and cost, schedule, and constructability evaluated as a part of the design.

13.6 OTHER CONSIDERATIONS

13.6.1 Replacement of Metallic Expansion Joints

Prior to replacement of metallic expansion joints by fabric joints, it is critical that the structural system and stability of the existing ductwork section are properly

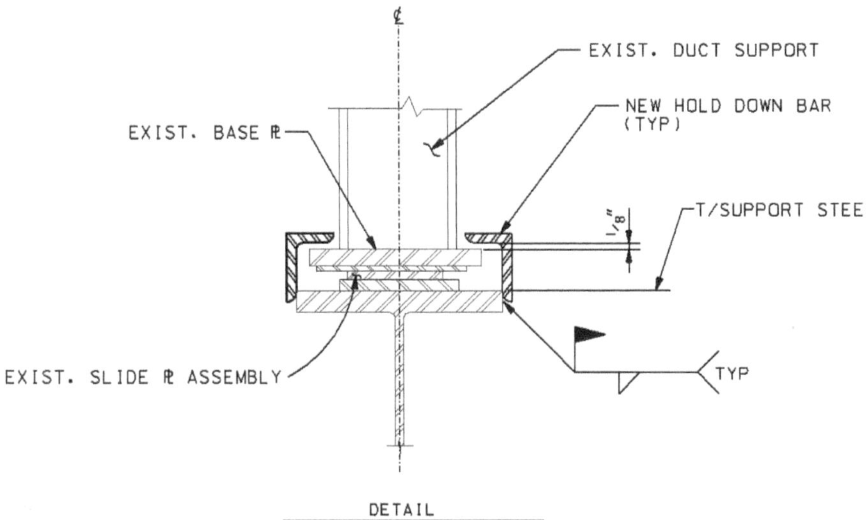

Figure 13-8. Example new inverted-L hold-down bars.

understood. In the past, some duct designs relied on the stiffness of the metallic expansion joints to provide lateral or vertical support for the duct section. In some designs, pairs of metallic joints are used with pin arrangements to create a "toggle section" that allows a lateral offset to occur (see Figure 1-7). Internal shear key arrangements may be included to transfer vertical or horizontal forces across the metallic joint opening. Replacing metal expansion joints with fabric joints may cause loss of support or stability of the existing ductwork. The lateral or vertical support should be appropriately replaced prior to removal of the metallic expansion joints.

13.6.2 Higher Flow and Flow Velocities

New design conditions and environmental controls may increase the volume of flow through an existing duct, which will increase the velocity of the flow. In these cases, sacrificial angles, flat plates, bent plates, or channels may be considered to provide erosion protection to the existing ductwork internal components. Figure 13-9 shows example details of sacrificial elements attached to an existing truss pipe.

Higher flow rates and flow velocities may also increase the vortex shedding forces on internals, and the internals may require reevaluation. Internals that are inadequate for the higher vortex shedding forces may need to be modified or replaced. Initial studies have shown that flat-sided ovals have a lower coefficient of friction than circular cross sections, so less flow eddies develop on the trailing edge. Internals can be modified to a flat-sided oval by welding plates and a half of a pipe to an existing truss member (Figure 13-9d).

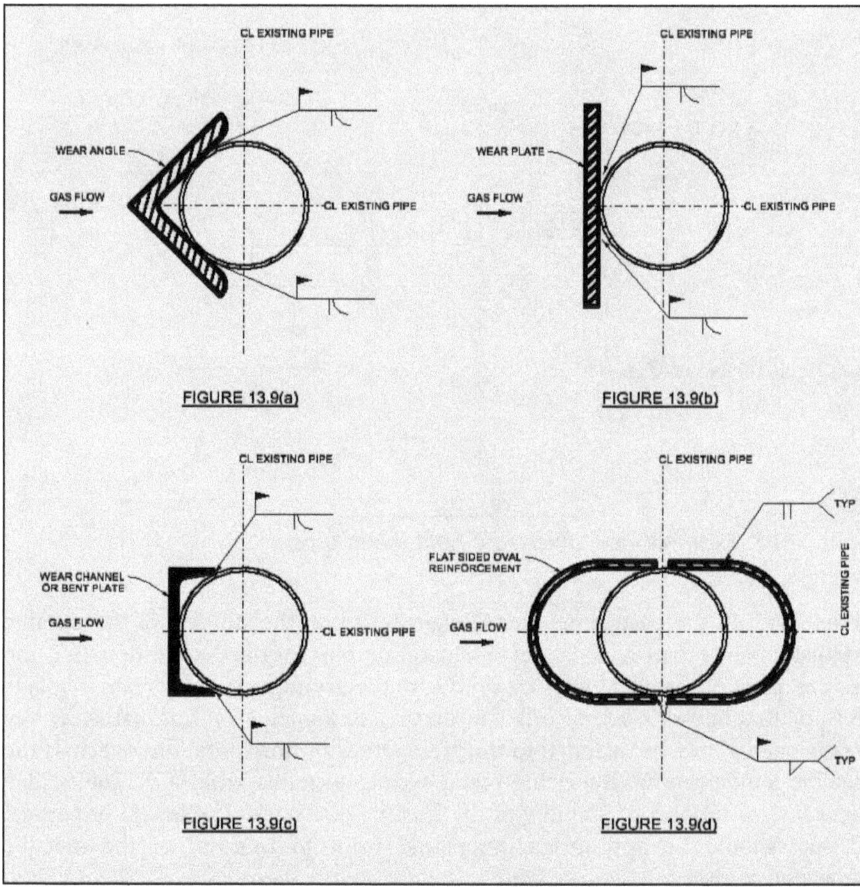

Figure 13-9. Example erosion protection for existing truss pipes and struts.

References

AISC. 2017a. *Specification for structural steel buildings.* AISC 360-16. Chicago: AISC.

ICC (International Code Council). 2017. *International existing building code.* IEBC 2018. Country Club Hills, IL: ICC.

Index

Equations are indicated by *e*; figures are indicated by *f*; tables are indicated by *t*.